D0919127

PROGRESS IN
PARASITOLOGY

UNIVERSITY OF LONDON
HEATH CLARK LECTURES 1968
delivered at
The London School of Hygiene and Tropical Medicine

Progress in
Parasitology

by

P. C. C. GARNHAM

C.M.G., M.D., D.Sc., F.R.C.P., F.R.S.

Emeritus Professor of Medical Protozoology, University of London
Senior Research Fellow, Imperial College of Science and Technology
Late Director, Department of Parasitology
London School of Hygiene and Tropical Medicine

UNIVERSITY OF LONDON
THE ATHLONE PRESS
1971

RECEIVED

JUN 23 1975

MANKATO STATE COLLEGE LIBRARY
MANKATO, MN

Published by
THE ATHLONE PRESS
UNIVERSITY OF LONDON
at 2 Gower Street, London WC1

Distributed by Tiptree Book Services Ltd
Tiptree, Essex

U.S.A.
Oxford University Press Inc
New York

© *University of London* 1971

ISBN 0 485 26321 1

QL757
.G37

Printed in Great Britain by
WESTERN PRINTING SERVICES LTD
BRISTOL

TO
MY WIFE
WHO ACCOMPANIED ME
ON MUCH OF THE PROGRESS

RECEIVED

JUN 2 3 1975

MANKATO STATE COLLEGE LIBRARY
MANKATO, MN

PREFACE

CHARLES HEATH CLARK (1860–1926) made his money in the tea-trade and later in the rubber plantation industry in Ceylon and South-east Asia. From these pursuits stemmed his interest in the tropics and he became a founder member of the League of Nations Union. Although his connection with the tropics was never more than second-hand, he was a philanthropist on a wide scale and devoted to service to mankind.

In the course of the last twenty years I have attended most of the Heath Clark lectures of the University of London and have come to realize that nearly half have depicted the landscape of disease. In this way the lecturer has complied with the terms of the Trust which require him to present, in a not too technical manner, the history, development and progress of preventive medicine and tropical hygiene in relation to social evolution in temperate and tropical climates. In the present series, I have tried to follow a similar course, understanding that development and progress entail comprehension of the aetiology of the disease and the elucidation of the life cycles of their agents.

I am deeply honoured by Sir Harold Himsworth, F.R.S. for taking the chair at these lectures, and to Dr E. T. C. Spooner, C.M.G., Dean of the London School of Hygiene and Tropical Medicine for inviting me to give them. Sir Harold has become familiar with tropical medicine during the last twenty years; I am nervous speaking in his presence, knowing that he will spot immediately any inaccuracy, superficial treatment or too hasty conclusion; the Dean is only too aware of my deficiencies, but I can rely on his characteristic generosity to ignore them now as he has done in the past decade.

The Dean gave me the title of these Lectures; but I must correct any misapprehension of their scope. *Progress* in Parasitology is not intended to be a résumé of the important series of annual volumes of *Advances* in Parasitology, which have been so ably edited by Ben Dawes (1963–1969). The word 'progress' is

371849

meant to imply a journey or expedition in the archaic sense as given in the Shorter Oxford English Dictionary; in other words, the subject is more the progress of the author than the subject, though an attempt is made to synthesize his experiences into something general. There is however a background of the past, and a foreground of the possible future.

I acknowledge with much gratitude the hospitality of the Rockefeller Foundation (itself closely involved in the history of the School and of innumerable public health enterprises in the tropics) for allowing me to write up these lectures in book form at its Villa Serbelloni at Bellagio on Lake Como. I am grateful to Dr C. A. Hoare, F.R.S., Professor A. Corradetti, Professor L. Bruce-Chwatt and others for various foreign references. I am indebted to countless colleagues in many parts of the world for practical help in the field and for their stimulating discussions about our subject; in particular to the late Dr R. B. Heisch, O.B.E., for his electrifying companionship.

Personal details of characters in our story have been most kindly supplied by Ambassador Carlos Chagas, Signora Grassi-Scola, Frau Elizabeth Bursy-Schaudinn, Mme Petrishcheva, Dr Clinton Manson-Bahr, the Marchese Castellani, Professor Claude Dolman, and many others, while Mr V. Glanville, Librarian of the School, gave me references to certain biographical details.

Permission was kindly granted by the Royal Society of Tropical Medicine and Hygiene for extensive excerpts from their Transactions; literary quotations are largely from out-of-date books or the author's own translations.

P.C.C.G.

CONTENTS

LIST OF TEXT FIGURES

LIST OF PLATES

ERRATA

The Addenda will be found on p. 205 not p. 200 as stated.

page 120 *line* 24   *at end of line read footnote reference* [1]

page 120 *note* 1   *for* Hutchinson *read* Hutchison

for Robert Knock *read* Robert Koch

I

INTRODUCTION

LIKE any other branch of science, parasitology has grown in recent years to monstrous proportions, and today nobody would be capable of producing a definitive work on the subject. The last textbook on a cosmopolitan basis, by a single author, was probably Brumpt's Précis de Parasitologie, the sixth and last edition of which was published in 1949; other books on parasitology of a more local interest continue to appear, such as those of Pessôa (1967) largely directed to Latin America, Chatterjee (1952) to India, and Jírovec (1960) to Europe, and although all give descriptions of the major parasitic conditions, the writers confine themselves to human disease, chiefly of the regions with which they are concerned.

The present volume is based on the Heath Clark lectures of 1968 and presents a few aspects of the subject with which the writer has come into contact during nearly half a century; it is thus very much drawn from his own experiences, half in the tropics and half in Europe. Examples are taken from both places and general, if speculative, principles have been deduced from them.

Emphasis has been placed on the natural history of infections and their relation to the environment. Throughout the book, the writer has tried to show not only the link between the parasite, its different hosts and the surroundings, but also the link between these and the parasitologist himself; just as Stanislavsky (1949) used to take every factor into account in order to expose the complete meaning of a play, or the surrealist painter or writer to add more and more to 'reality', so in parasitology is it desirable to take the widest view in order to obtain an idea of the true composition of the subject. Thus in the later chapters, problems of both parasite and parasitologist are considered, and finally the life and works of some typical parasitologists of the past are studied in relation to their environment, and not only the physical reality, but as seen through an artist's eye.

There is a tendency to regard parasitology as little more than a branch of natural history or, at most, as a minor handmaiden of zoology, whose major interests (Weiss, 1968) have shifted to molecular biology, immunology, biochemistry and genetics. There has been a brain drain from parasitology to these modern sciences, but the traffic is beginning to move in the opposite direction, and immunologists like John Humphrey, Robin Coombs, Howard Goodman and Dumonde are mingling in our ranks; geneticists like Beale have become interested, while biochemists such as von Brandt, Moulder and James Williamson have long been involved. Parasitology needs these new disciplines as much as any other branch of biology. Our subject itself is fundamental to human life: we are parasites from conception to the grave—parasites on our mothers in prenatal life, and on nature in post-natal. The general principles which emerge in a study of the formal subject are often applicable to life in general and to society in particular.

The scope of parasitology has been much debated and there is no general agreement about what should be included. In 1835 Bassi, the pupil of Spallanzani of Pavia, wrote in the introduction to his book on silkworm disease that all infectious diseases of man, animals and plants had a parasitic origin. This idea became firmly established by the time of Pasteur and Koch, and all macro- or micro-organisms living inside or on the surface of the host were regarded as parasites. In this way, arthropods, helminths, protozoa, fungi, bacteria, rickettsiae, spirochaetes and viruses, in time became included and Brumpt's (1949) classic work is devoted to all these pathogenic agents. This has continued to be the basis of the parasitology of most Latin authors and, as recently as 1966, the editor (Guerra) of the *Cuban Review of Tropical Medicine* emphasized the need of uniting the subjects of entomology, microbiology and parasitology in a single department and deplored any amputation. Burnet (1966) also in writing about the natural history of infectious disease makes it quite clear that he regards all infections as parasitic 'be it virus, bacterium, worm or insect', and in order to limit his study to reasonable proportions he still included bacteria, viruses and protozoa in his field.

Anglo-Saxon practice is to separate these disciplines, and the usual textbook on parasitology includes only protozoology and helminthology; entomology is largely discarded and even the two former subjects may become split into different departments in an institute where it is not uncommon to find the parasitologist identified solely with the helminthologist. Of course the virologist needs separate laboratories, but when divorced from the microbiologist, he fails to recognize an *Acanthamoeba* in his cultures and ascribes the peculiar effects of this protozoon to the so-called Ryan virus. Or the protozoologist needs the help of the helminthologist to work out the life cycle of *Histomonas* in the nematode worm—*Heterakis*; while the parasitologist is always hand-in-glove with the entomologist. Moreover, the general principles of parasitology *sensu latu* cannot be appreciated by the narrow specialist. Levine (1965) in a discussion on veterinary parasitology states that, without the whole subject of microbiology, the parasitologist becomes too limited, and like a horse with blinkers, his field of vision is restricted.

It is perhaps significant that the first International Congress of Parasitology was held in *Rome* in 1964, and that the wider aspects of the subject were included in its proceedings, whereas the Second Congress held in *Washington* in 1970, confined itself largely to protozoology and helminthology.

In order to prevent too wide a diffusion of parasitology it is useful to impose some sort of restriction, and this perhaps can best be done by adding to the classical subjects of protozoology and helminthology, only *vector-borne infections* of other types, like plague, or blue-tongue of sheep, but not diphtheria, or rinderpest. In this way, principles emerge which are generally applicable, and this is largely the scope of this book.

REFERENCES

BASSI, A. (1835) *Del mal del segno, calcinaccio o moscardino, malattia che affligge i bachi da seta e sul modo di liberarne le bigattiere anche piu infeste*, Lodi.
BRUMPT. E. (1949) *Précis de Parasitologie*, 6th ed., Paris: Masson.

BURNET, M. (1966) *Natural History of Infectious Disease*, Cambridge: University Press.

CHATTERJEE, K. D. (1952) *Human Parasites and Parasitic Diseases*, Calcutta: Sree Saraswaty Press.

GUERR, F. S. (1966) 'Tendencias modernas de la parasitologia', *Rev. Cub. Med. trop.* **18**, 81–82.

INTERNATIONAL CONGRESS OF PARASITOLOGY I (1966) Rome, 2 vols., Milan: Pergamon.

INTERNATIONAL CONGRESS OF PARASITOLOGY II (1970) Washington, 3 vols., *Suppl. J. Parasit.*

JÍROVEC, O. (1960) *Parasitologie für Ärtze*, Jena: Gustav Fischer.

LEVINE, N. D. (1965) 'Parasitology', *Am. J. vet. Med.* **26**, 434–43.

PESSOA, S. B. (1967) *Parasitologia Medica*, 7th ed., Rio de Janeiro: Koogan.

STANISLAVSKY, K. (1949) *Building a character*, New York: Theatre Art Books.

WEISS, P. A. (1968) *Dynamics of Development: Experiments and Inferences*, New York: Academic Press.

2

GENERAL PRINCIPLES OF
THE ZOONOSES

PARASITOLOGY can be regarded, in many respects, as a study of the zoonoses, using the latter word in a broad sense and by dint of stretching the point here and there. The word 'zoonosis' merely indicates a disease of animals but this is about the only meaning in which it is rarely or never used today. The common attribution to Virchow of this term is apocryphal, and the history of its introduction into scientific writing is discussed in interesting detail in the monograph by Fiennes (1967) on simian diseases in relation to man. Many definitions have been given, but the simplest is that of the World Health Organization (1967) which reads 'those diseases and infections which are naturally transmitted between vertebrate animals and man'. This definition is much too anthropomorphic in the context of parasitology, and a wider approach was suggested by the writer (Garnham, 1969) who divided the zoonoses into two categories, medical (as defined above) and veterinary zoonoses by which are meant infections which are naturally transmitted between wild vertebrate animals and domestic vertebrate animals. But the latter definition still retains an anthropomorphic tinge, in that domestic animals imply animals domesticated by man, and a still broader outlook is necessary.

A useful classification was supplied by Kozar (1959) who placed the parasites into three groups as follows:

1. Zooparasites
   (a) Absolute—highly specific to animals, and man completely insusceptible (the true zoonoses).
   (b) Relative—less specific to animals, and man occasionally susceptible to the larvae or even the mature forms (the relative anthropozoonoses).

2. Anthropozooparasites
   The parasites show an equal specificity to man and animals
   (the anthropozoonoses proper).
3. Anthropoparasites
   (a) Absolute—highly specific to man, and animals completely insusceptible.
   (b) Relative—less specific to man, and animals occasionally susceptible.

A further term 'amphixenosis' was introduced by Hoare
(1962) to indicate infections which are maintained in both man
and animals and are transmitted in either direction.

The writer (Garnham, 1958) suggested that the zoonoses
comprised two groups, in one of which man is an *essential* link in
the process—the so-called 'euzoonosis', and the other in which
man is only accidentally involved—the so-called 'parazoonosis'.
The former is a rare condition and is represented by the tapeworms, *Taenia saginata* and *T. solium*; the latter includes the vast
majority of infections of all aetiological types.

Probably all the communicable diseases of man were at one
time or another zoonoses. Some remain so, like rabies; others
scarcely retain any link with animals, for instance, influenza or
malaria; while most infections lie in between these extremes, and
under different epidemiological circumstances may be entirely
zoonotic, like the Central American form of leishmaniasis, or
entirely human, like the Indian form of this disease.

It is possible to discuss the zoonoses in two ways, starting
either with the animal reservoir or with the infective agent. One
might take the dog as an example and show how this animal
carries a great variety of infections, including rabies, South
American trypanosomiasis, leishmaniasis, toxoplasmosis and
many helminths; or cattle which harbour the agents of anthrax,
Rift Valley fever, tapeworm, Q fever, toxoplasmosis and piroplasmosis—all of which infections they pass in different ways to
man. Innumerable other animals could equally well be cited,
from armadillos to parrots, monkeys, rodents, and fish. Of them
all, rodents are probably most widespread and numerous, and
these animals constitute the most important *primary* reservoirs of
infection, while dogs are the most significant secondary reser-

voirs. The primary reservoir is the natural or feral host; the secondary (in the case of the dog) is usually a liaison host which brings the infection into contact with man.

But the second approach is more rewarding, for it is possible to consider some typical infective agents and trace their zoonotic pathway from the original wild animal host to man and his domestic animals, until the infection becomes finally fixed in man and the zoonosis as such is extinguished. Examples are given in the next chapter from amongst the protozoa, bacteria, viruses, spirochaetes and rickettsia. A similar course of evolution is apparent in helminthic infections, but as a rule, the progress is more limited and is governed by a different set of laws; as metazoan parasites they are too large and structurally differentiated to participate in the complete process; e.g. when vector transmitted, they are absolutely dependent upon this method of transmission and direct passage of a worm from man to man is impossible. Another important difference between the course of infections due to micro-organisms and metazoans respectively is that the latter do not usually multiply in the host; disease is only produced when many helminths are introduced.

For these reasons, the generalizations applicable to micro-organisms are usually irrelevant to helminths; protozoan parasites, though belonging to the Animal Kingdom as do the helminths, behave much more like bacteria, rickettsiae and fungi which are of a plant nature. On the other hand, some of the best examples of zoonoses are to be found in infections due to trematodes and nematodes and a few of these are discussed in Chapter 5.

Four major principles can be drawn from a study of the zoonoses, as follows:

1. The natural focus, representing the scene of occurrence of the zoonosis.
2. The dynamic or evolutionary pattern of the zoonosis.
3. The contrasting severity of the infection in natural and unnatural hosts.
4. The best reservoir is the partially resistant animal.

THE NATURAL FOCUS OR NIDALITY OF INFECTION

The zoonosis occurs in a natural focus and is characterized by a special landscape epidemiology. This concept was developed by the Russian School of Pavlovsky (1966) as the result of observations made particularly on spring-summer encephalitis, leishmaniasis, tularaemia, leptospirosis, rickettsiosis, Q fever, brucellosis, and plague, between the two world wars. The theories were quickly applied in Poland and Czechoslovakia, and by Karl Meyer in California, Heisch in East Africa and Audy in Malaysia. Audy (1968) described the ecology of scrub typhus on this basis in the Heath Clark lectures of 1965. The Pavlovskian doctrine was undoubtedly based on earlier investigations on the natural history of disease, e.g. the work of the French School in North Africa (Charles Nicolle in Tunisia and Georges Blanc in Morocco), and of the Rocky Mountain spotted fever laboratory in Montana, on plague by Wu Lien Teh (1936) in Manchuria, and on trypanosomiasis by Swynnerton (1936) in tropical Africa. But the interest of the Russian work lies in its synthesis of a wide variety of zoonotic infections into general principles, a task which Nicolle (1930) alone of the above had attempted earlier in the century.

The natural focus occurs in a characteristic landscape and this is best studied by taking a bird's eye view of the locality concerned. One starts by looking at the human disease, examining patients in hospitals and by surveys of the population; but equal attention is paid to the animals which may constitute the reservoir of infection, to the trees, burrows or caves which shelter the animals, and to the arthropods which bite them. Then the food and habits of all these creatures, including man himself must be ascertained. A study is also made of the geology, rainfall, humidity, and winds of the place, until finally the time comes when one becomes so familiar with the general aspects that the investigator can go to another place and recognize from a glance at the landscape if a similar pattern exists there, to spot where the danger lies and where the infection is likely to be contracted or perhaps where it can be guaranteed to be absent.

Pavlovsky's holistic approach is closely related to Southwood's (1969) 'Components of Population Systems' which include five main elements: climate, food, space in which to live, the influence of other animals on population density and the population under consideration itself; these interact with each other in complex ways which Southwood, unlike the Russians, was able to express in mathematical formulae.

Medical geography is yet another aspect of the subject and Dudley Stamp spoke about it in the Heath Clark lectures of 1962. He emphasized in particular the effect of climate on the incidence of infective diseases and showed that, as a whole, their greatest prevalence is in the tropics. Unfortunately, we are ignorant of many of the factors; 'we know more about the effect of climate on fruit trees than on human beings!'

However, medical geography omits zoogeography and the latter is of course an essential element of the nidality of disease. Hoogstraal (1956) who has spent a lifetime on this subject, states that we must study the biology and ecology, not just of the obvious members of the local fauna, but of the whole range; faunal exploration on a wide scale is the basis of any scientific approach to the zoonoses.

But even zoogeography does not go far enough and the doctrine of the nidality of disease should be based on *biogeography* in order to produce full results. This approach is being attempted by Fittkau *et al.* (1968) in relation to South America, where the relatively recent colonization by man has produced situations of great interest to students of the zoonoses, as will be indicated in later chapters of this book.

DYNAMIC PATTERN OF ZOONOSES

The course of an infection is never static but changes from one day, year or century to the next and the pattern is particularly susceptible to alteration with the advent of man. This occurred when man was evolving from the prehominids and exchanged his sylvatic environment for the savannah. Lambrecht (1968) attempted to trace the incidence of human infections in relation to geological ages and suggested that in the Tertiary period the arboreal prehominids were exposed to yellow fever and malaria; in the Pleistocene period primitive man was brought into contact with trypanosomiasis and leishmaniasis; and in the Upper

Pleistocene period cave-dwelling man acquired rickettsial and helminthic infections to be followed by bacterial and viral diseases.

During the embryonic stage of the zoonosis, man is absent, and the infection is confined to its natural host. But with the intrusion of man, the situation becomes unstable, changes are at once initiated, new reservoirs may become involved like domestic animals, different species of arthropod vectors transmit the infection and, with increasing civilization, the animals retreat, the vector fades from the picture and the disease rages by direct interhuman contact. The aetiological agent may become transformed during this cycle of events and acquires different antigenic properties, though it is difficult or impossible to prove that a new species of organism has arisen in this way.

In the course of evolution, the zoonotic aspect of an infection may completely disappear as in most forms of malaria and Gambian sleeping-sickness; at the other end of the scale lie the unborn zoonoses waiting their opportunity to become established in man; in between these extremes are found the zoonoses which are still evolving, like rickettsial infections, arboviruses and leishmaniasis.

These changes do not necessarily occur only in one direction; sometimes a zoonosis goes into reverse, and this is best seen in recent epochs when large scale human migrations have taken place. Emigration to the New World was accompanied by the introduction of infections like yellow fever, leishmaniasis and possibly malaria which spread back into the indigenous wild animal population. Similarly, plague reverted to the primary 'sylvatic' form in new foci in South Africa, the United States, and possibly South America by the arrival at the ports of secondary rodent reservoirs. Some of these episodes are described in the following chapter.

VARIATION IN THE SEVERITY OF INFECTIONS
IN THE NATURAL AND UNNATURAL HOST

The disease is usually least virulent when it is acquired directly from wild animals, but when the organism becomes thoroughly adapted to man, it multiplies more readily (perhaps by a variation in its iso-enzyme patterns) and the mortality of the infection is much heightened. Feral strains (i.e. strains derived from a wild animal source) are feebly pathogenic; human strains are much more virulent. The increased virulence may possibly be merely a reflection of a higher inoculum of the infective agent, with a small dose, the immunity mechanism of the animal has time to become established, with a bigger dose, the infection gets the upper hand.

There are exceptions to this generalization, such as rabies, but the transference of a feral infection to any new host is usually effected with difficulty, owing to the lack of virulence of the organism in the different environment. Various examples are given later, but human toxoplasmosis provides one of the best, and is mentioned here as an illustration. This infection is very mild or inapparent in the human population which probably acquires it from a zoonotic source. If, however, the organism is passed via the placenta from mother to foetus, the infection often becomes rampant and kills the foetus *in utero* or soon after birth. Equally, infections acquired by man in laboratories from unnatural hosts are usually severe. Feral strains of *Toxoplasma gondii* are often of such feeble virulence, that they must be blind-passaged in the new host for a number of times before they are rendered apparent.

When simian malaria parasites are inoculated into man, at first they find difficulty in maintaining themselves; however, after several passages from man to man, the infection may become virulent. Thus, after Ciuca *et al.* (1955) had passed *Plasmodium knowlesi* a number of times through their patients suffering from cerebral syphilis, the parasitaemia reached such a height (over half a million organisms per ml) and the fever became so severe that they had to abandon this form of therapy.

David Lewis, Nadim, and the writer recently isolated in culture, strains of *Leishmania tropica* from *Phlebotomus papatasii* and *P. caucasicus*, but completely failed to establish them in mice or hamsters; the strains had originated from infections in the wild host, *Rhombomys opimus*, and as had been noted by Nadim and Faghih (1968) such feral strains always possess a low degree of virulence for exotic hosts.

BEST RESERVOIRS

Stability of a zoonosis is achieved when the wild animal host is relatively resistant to the infection. Yellow fever is maintained in an almost unrecognized form in the monkeys of tropical Africa, because these animals rarely die from the infection; but in Latin America, the monkeys belong to a completely different type, they are highly susceptible to the virus and epizootics of great violence sweep through the forests and the zoonosis becomes extinguished because the reservoir has been destroyed.

Baltazard *et al.* (1952) pointed out in Persia and Heisch *et al.* (1953) in East Africa, that the wild rodents of these places are fairly resistant to *Pasteurella pestis* which can persist for a long time, giving rise to sporadic cases of 'sylvatic' plague in man; if the infection however spreads to the susceptible domestic rats, fatal epizootics in these rodents occur and severe epidemics afflict man. But this type of plague is self-limiting owing to the exhaustion of the supply of the domestic reservoir; and the process comes to an end.

The relevance of these general principles of the zoonoses is indicated in the examples given in the following chapter.

REFERENCES

AUDY, J. R. (1968) *Red Mites and Typhus* (Heath Clark Lectures, 1965) University of London, Athlone Press.

BALTAZARD, M., BAHMANYAR, M., MOFIDI, C. and SEYDIAN, B. (1952) 'Foci of plague in Kurdistan', *Bull. Wld Hlth Org.* **5**, 441–72.

BRUMPT, E. (1949) *Précis de Parasitologie*, 6th ed., Paris: Masson.

CIUCA, M., CHELARESCU, M., SOFLETEA, A., CONSTANTINESCO, P., TERITEANU, E., CORTEZ, P., BALANOVSCHI, G. and ILIES, M. (1955) 'Contribution expérimentale à l'étude de l'immunité dans le paludisme', *Editura Acad. Rep. Pop. Romane.*

FIENNES, R. (1967) *Zoonoses of Primates*, London: Weidenfeld and Nicholson.

FITTKAU, E. J., ILLIES, J., KLINGE, H., SCHWABE, G. H. and SIOTI, H. (1968) *Biogeography and Ecology in South America*, vol. 1, The Hague: W. Junk.

GARNHAM, P. C. C. (1958) 'Zoonoses or infections common to man and animals', *J. trop. Med. Hyg.* **61**, 92–94.

GARNHAM, P. C. C. (1969) 'Malaria parasites as medical and veterinary zoonoses', *Bull. Soc. Path. exot.*

HOARE, C. A. (1962) 'Reservoir hosts and natural foci of human protozoal infections', *Acta Trop.* **19**, 281–317.

HOOGSTRAAL, H. (1946) 'Faunal exploration as a basic approach for studying infections common to man and animals', *East Afr. med. J.* **33**, 117–74.

KOZAR, Z. (1959) 'The problem of parasitic anthropozoonoses', *Wiadomości Parazytologieczne* Warsaw **5**, 199–211. (Fuller version in Polish 175–98.)

LAMBRECHT, F. L. (1968) 'Notions concerning the evolution of communicable diseases in man', *South Afr. J. Sci.* **64**, 64–71.

NADIM, A. and FAGHIH, M. (1968) 'The epidemiology of cutaneous leishmaniasis in the Isfahan province of Iran', *Trans. R. Soc. trop. Med. Hyg.* **61**, 534–42.

NICOLLE, C. (1930) 'Destins des maladies infectieuses', reprinted *Arch. Inst. Pasteur Tunis* 1956 **33**, 1–179.

PAVLOVSKY, E. M. (1966) *Natural Nidality of Transmissible Disease*, ed. Norman D. Levine, Urbana: University of Illinois Press.

SOUTHWOOD, T. R. E. (1969) 'The abundance of animals', *Inaug. Lect. Imp. Coll. Sci. Technol.* **8**, 1–16.

STAMP, D. (1962) *Some aspects of medical geography* (Heath Clark Lectures 1962) University of London, Athlone Press.

SWYNNERTON, F. C. M. (1936) *The Tsetse Flies of East Africa, Trans. R. ent. Soc. London*, **84**.

TEH, W. L., CHUN, J. W. H., POLLITZER, R. and WU, C. Y. (1936) *Plague*, Shanghai: National Quarantine Service.

WORLD HEALTH ORGANIZATION (1967) Joint FAO/WHO Expert Committee on Zoonoses: third report, Technical Report Series no. 378, Geneva.

3

EXAMPLES OF ZOONOSES

Pavlovsky (1966) stated that his theory of 'natural focality' was based on the following types of disease: viral, rickettsial, spirochaetal, bacterial, protozoal, helminthic, arthropod, leptospiral, and mycotic. This chapter and the two following are concerned with examples from all but the last two categories in this list, and as zoonoses they represent illustrations of his specific outlook.

PROTOZOAL INFECTIONS

The majority of parasitic protozoans represent zoonoses of one kind or another, and three examples are taken here to illustrate the process in full operation. In the following chapter the malaria parasites are considered as a special case, and elsewhere in this book, reference is made to interesting organisms such as piroplasms, *Balantidium coli*, *Entamoeba histolytica*, *Toxoplasma gondii* and others.

Leishmaniasis

It is misleading to use the single term 'leishmaniasis' for the numerous conditions whose zoonotic aspects are now to be described, for these infections, although due to organisms with an identical morphology, in all other respects are as different from each other as chalk from cheese. The problem can be approached in one of two ways: all the 'strains' can be unified into one species (under the name of *Leishmania donovani*) or separate specific status can be given to each strain. The former solution is taxonomically correct, but the latter appeals to the epidemiologist who likes to have a distinctive label available for the organisms with which he is concerned, and will be the procedure largely used here. Speciation has however flowered so profusely in Latin America that nomenclature has scarcely kept pace, and several characteristic epidemiological types remain anonymous or are referred to by the general name, *L. braziliensis sensu latu.*

The disease has a cosmopolitan distribution, occurring in temperate as well as tropical zones; the infection spreads as far north as the suburbs of Paris, and one of the vectors, *Phlebotomus perniciosus*, even to the Channel Islands. There are two main forms of leishmaniasis, cutaneous and visceral, as well as innumerable varieties, for the changing environment in different parts of the world is conducive to speciation in the causative organism.

The condition or conditions represent the ideal illustration for the theories advanced in the previous chapter, and it is essential to approach the problem from the pristine angle. The mammalian species of *Leishmania* all stem from two main groups, as follows:

1. Species originating in wild Canidae with final hosts in dogs and man.
2. Species originating in wild rodents, with final hosts in dogs and man.

The origin of the mammalian species is necessarily obscure, though Hoare (1967) suggested that lizards were probably the original vertebrate hosts of *Leishmania*; lizards in the Old World (though never in the New) are frequently found to be infected today. One species, *L. adleri*, discovered by Heisch in Kenya, was shown by Adler and Adler (1955) to have immunological affinities with the human species, and to give rise to feeble infections in man. Various carnivorous animals feed on lizards, and today the major animal reservoirs of the cutaneous infection are jackals, foxes and other wild canines. The lesions on the rodents are usually cutaneous, and in jackals, etc., visceral, though the distinction is not absolute.

Lizard species of *Leishmania* may have some significance in regard to the incidence of the human disease; Southgate (1967) discovered that experimental infections with *L. adleri* confers immunity against ground-squirrel strains of *L. donovani* and suggested that such an effect might occur in nature. Part of a population in an endemic area might thus be protected by this

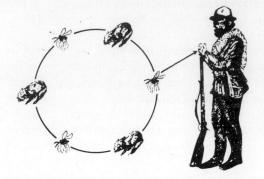

FIG. 1. Progress of zoonosis: hunter intrudes into cycle between wild animals and vector. (Figs. 1–5 from Garnham 1963.)

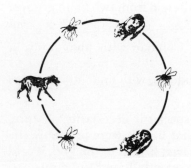

FIG. 2. Progress of zoonosis: dog intrudes into cycle between wild animal and vector.

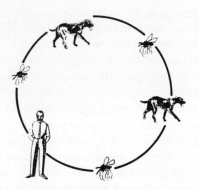

FIG. 3. Progress of zoonosis: domestic cycle of dog, vector and man.

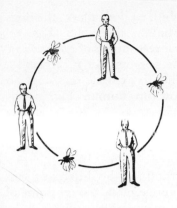

FIG. 4. Progress of zoonosis: domestic cycle between man and vector (extinction of zoonosis).

FIG. 5. Progress of zoonosis: domestic cycle with loss of vector (congenital transmission).

accidental vaccination via the bite of *Phlebotomus* infected with the harmless lizard parasite.

The general progress of the zoonosis is thought to take the following form and is applicable to all varieties of the disease.

The primeval infection is confined to a specific wild animal and its special *Phlebotomus* vector or vectors; man is outside this cycle of transmission but, sooner or later (Fig. 1), he intrudes into the feral environment as a hunter, for getting firewood, or for collecting honey, chicle or other products of the forest or steppe. Dogs may accompany their masters and become infected (Fig. 2). Sporadic infections may then occur without disturbing the even course of the infection; but the territory may be attractive, and man occupies it first in primitive settlements and later more permanently. The environment becomes profoundly altered; the wild animals disappear but, in their place, the dogs or other domestic animals act as liaison reservoirs (Fig. 3), while instead of burrow-haunting or forest species of *Phlebotomus*, transmission is effected by domestic vectors like *P. argentipes*. Domiciliation of the infection becomes so extreme that dogs no longer play a part and the cycle is now interhuman (Fig. 4), but still initiated by the bite of the *Phlebotomus*. At last, even the insect may vanish and humanization becomes complete so that airborne, venereal or transplacental infection replaces the natural route (Fig. 5).

It is possible to trace the biogeography of the different forms of leishmaniasis on the basis of the above plan, though some of the links in the chain are inevitably speculative.

Probably the oldest focus of the cutaneous form is the Kara Kum desert of Turkmenia and the adjoining plateaux of northern and central Iran. To the north and east, the natural focus spread into the Turanian steppes (e.g. Kazakstan) and eventually reached Mongolia, along the routes which Marco Polo (1968) took in the fourteenth century. Hindle and Thomson (1928) studied the disease in North China and demonstrated the excellence of Mongolian hamsters as hosts of the organism. South-east of the original focus, the infection entered Pakistan and petered out in India.

In the Russian steppes, the gerbil (*Rhombomys opimus*) reigns supreme (Plate I, Fig. 6); it lives in burrows, *Phlebotomus* feeds on it, and the microclimate is ideal for the transmission of *Leishmania*. Human cases only occur by accident during hunting expeditions, but the hunter may bring the infection into his village and set up the next stage of the zoonosis.

The natural foci of leishmaniasis in Iran are precisely similar to those in southern Russia, but the territory has been occupied by man for at least two thousand years in some parts, like the region of Isfahan; the villages are surrounded by gerbil burrows and there is close contact between man and animal. In fact, the gerbils are almost part of the village; they prefer to live where the subsoil water is not much deeper than 15 m and this is what suits the villagers also. The gerbils play around the entrances to their burrows in the evening; the children are close by, and the *Phlebotomus* bite both in their hundreds. In such villages, like Aliabad 25 km to the east of Isfahan, the infection rate in the gerbils (where the lesions are chiefly on the outer ear) is 40 per cent and in the children 100 per cent (Nadim and Faghih, 1968), while in sandflies (*P. papatasii* and *P. caucasicus*) which David Lewis, Nadim and the writer collected near this village in September 1968, flagellates were found to be present in the gut of 20 per cent. Only young children showed active lesions, but all the older inhabitants of Aliabad had typical scars from leishmanial ulcers. Some babies only 2 weeks old had ulcers, because this was the 'wet form' of the disease due to *Leishmania tropica tropica* (= *major*) of zoonotic origin, unlike the more chronic 'dry form' due to *L. t. minor* of the towns. The lesions in the children are not severe, perhaps because some cell mediated immunity is inherited from the immune mothers.

The larger the village, the further removed becomes the natural focus, because the sandfly has a limited flight range, usually not exceeding 250 m, while the animals are driven away, and in such small towns, the infection rate drops to 40 per cent. The infection is taken from these natural foci into the neighbouring towns, where the dogs become secondarily involved and the primary zoonosis disappears, as in Isfahan and Teheran. The species of *Phlebotomus* changes, *P. caucasicus* is left behind in the gerbil burrows, *P. papatasii* goes as far as the villages, but in the big cities, *P. sergenti* is the main vector in this region.

These steppes of Central Asia are thought by Lysenko (1967) to constitute the ancient source of visceral leishmaniasis (kala-azar); the endemic disease spread in the rural zones, secondarily infected dogs, and extended to towns in Transcaucasia, eventually reaching the Mediterranean to the west and India and China (via the Gobi Desert) to the east. Latyshev *et al.* (1947) incriminated the jackal as the wild animal reservoir in the Vakhsh Valley of south Tajikistan; later foxes, wolves and bears were found to be involved, either primarily or secondarily.

The distribution of kala-azar does not entirely overlap that of cutaneous leishmaniasis in this region; the visceral form occurs here at a rather higher altitude (up to 2300 m). The landscape epidemiology in Central Asia is in contrast to Ethiopia where the cutaneous disease occurs in the mountains and the visceral in the lowlands. Kala-azar is sporadic in nomads in Iran, but small

localized outbreaks may occur in children, as in the village of
Gah Ferok, about 50 km from Isfahan, where the vector is
probably *P. major*.

Under certain circumstances, but probably as the result of
high rainfall, kala-azar may become epidemic as in Bengal and
Assam; direct interhuman transmission occurs, and even the
dog no longer plays a role, probably because the vector species
of sandfly (*P. argentipes*) does not feed on dogs. This non-
zoonotic type of the disease was almost always fatal before its
curability with antimony was demonstrated by Sir Leonard
Rogers.

The normal pattern of visceral leishmaniasis is lost when the
causative organism no longer becomes transmitted by the insect
but directly from person to person. Congenital cases of kala-azar
were first demonstrated by Low and Cook (1926) and venereal
infections by Symmonds (1960); but such examples though
natural, are rare. Unnatural transmission by blood transfusion
has been reported from Sweden and England (Kostmann *et al.*,
1963).

The landscape epidemiology of cutaneous leishmaniasis is
similar in Turkey, Iraq, Jordan, Syria and Israel, though the
more distant from the ancient focus in Central Asia, the less
conspicuous is the wild animal reservoir. In Israel, however,
during the war with the United Arab Republic in 1967, hundreds of soldiers contracted the disease when camped in semi-
desert country, where the merions and other rodents were found
by Gunders *et al.* (1968) to be infected with *L. tropica*. A similar
terrain is found in North Africa on the slopes of the Atlas Mountains, and particularly in two places—the Gafsa and Biskra
oases—where the southern slopes descend to the Sahara
desert. Here the disease used to be so common that it went by
the name of Gafsa sore and Biskra button respectively. Today,
in Gafsa, the use of aerial DDT sprays against locusts has
fortuitously obliterated all the sandflies and the infection has
disappeared.

The dog is a secondary reservoir of both forms of leishmaniasis
in the Mediterranean subregion of the Palearctic, where a wild
animal reservoir is rarely encountered. However, Rioux *et al.*
(1967) found two foxes infected with *L. donovani* (? = *infantum*)

PLATE I

FIG. 6. In the Turkmenian Steppe. *Rhombomys opimus*, reservoir of *Leishmania tropica* (courtesy of D. J. Lewis).

FIG. 7. Acacia forest in the Southern Sudan: zoonotic focus of kala-azar and Rhodesian trypanosomiasis (courtesy of H. Hoogstraal).

PLATE II

FIG. 8. Termite hill in Northern Kenya: non-zoonotic focus of kala-azar (courtesy of R. B. Heisch).

FIG. 9. Devastation of forest in Brazil: conditions become ripe for epidemics of espundia.

on the southern slopes of the Massif Central and the eastern Pyrenees in France respectively, at altitudes of up to 500 m. Although widespread in southern Europe, including the Balkans, the visceral infections are rarely more than sporadic. Rioux (1969) associated a highly characteristic landscape epidemiology with visceral leishmaniasis in southern France; the infection is particularly prevalent in hilly country containing mixed forest of *Quercus ilex* and *Q. pubiscens* and accompanied by heavy breeding of *P. ariasi*. Is this perhaps a special feature of the natural focus of *L. infantum* as contrasted with the oriental and African *L. donovani?*

In Africa, the disease extends around the fringes of the Sahara, but as the territories become more tropical, the landscape epidemiology changes. In the southern Sudan, kala-azar is a serious problem, the organism has been found in various wild animals (by Hoogstraal and Dietlein, 1964) including the genet and scrval cats, mongoose and particularly various rodents (such as *Arvicanthis* and *Mastomys*). The biogeography of the condition here is presented in great detail in a monograph by Hoogstraal and Heyneman (1969). The investigations were made in the Paloich-Malakal districts lying at the eastern extremity of the Sudanese savannah belt which stretches across Africa between the Sahara and the equatorial rain forests to the south. In this vast zone, leishmaniasis assumes different forms and the observations of the American workers relate to a special landscape characterized by park-like Acacia-Balanites woodlands (Plate I, Fig. 7) which are flooded during the rainy season of June to October; in the dry season, the 'black cotton' soil becomes extensively cracked and deep crevices are thought to provide an ideal microclimate for the phlebotomine vector (*P. orientalis*) of the disease (Kirk and Lewis, 1947).

Several members of the American team who were investigating kala-azar in the Sudan, developed cutaneous leishmanial sores on the face which completely failed to visceralize (Cahill, 1964), just as in Malta, *L. donovani* sometimes fails to extend beyond the lymph nodes. Similar cutaneous and glandular cases in France and Spain were also thought by André *et al.* (1957) to be due to infections of *L. infantum* which fail to visceralize; the French workers suggest that inapparent leishmaniasis may be common in these endemic regions and that there is widespread immunity to kala-azar as a consequence. These responses are examples of the feeble pathogenicity of infections of zoonotic origin.

A similar picture is seen in Kenya, where rodent strains of *L. donovani* proved to be avirulent when inoculated into volunteers, and these men were shown by Manson-Bahr (1963) to be immune when challenged with the virulent organism. The infection in Kenya has been found in ground-squirrels and gerbils (*Tatera* spp.), and sporadic cases of kala-azar have been known for a long time in the (largely uninhabited) enzootic area, with many similarities to the Sudanese foci; but more recently, kala-azar became introduced into the populated district of Kitui, where the infection was spread in epidemic form from one person to another by the bite of peridomestic sandflies. These insects breed in holes in the termite hills (Plate II, Fig. 8) and emerge in the evening to bite the youths who sit around them. There is apparently no wild animal or canine reservoir in this locality (Heisch, 1963). Actually nowhere in tropical Africa have infections in dogs been demonstrated; perhaps the reason for this absence is the emaciated state of the 'pi-dog' in these regions, as it is well-known that leishmaniasis can only become established in well-nourished animals.

Cutaneous leishmaniasis extends into the highlands of Ethiopia and around the southern fringes of the Sahara in West Africa, but further south there is a barrier of unknown nature and the disease is practically absent below the Equator; even visceral leishmaniasis is confined in this continent to the northern hemisphere, although potential vectors and reservoirs are plentiful.

The landscape epidemiology of leishmaniasis in the New World is completely different from that of the Old World; it is associated with forest instead of semi-desert country and is clearly of indigenous origin because of the wide distribution of the animal reservoir. David Lewis and the writer visited in 1959 British Honduras and Mexico, territories representing the northern extremity of the disease, and soon realized that the infection must be a zoonosis, for the cases occurred solely in scattered groups of chicleros and mahogany workers who live for months on end in the depths of the forest. They were usually in twos or threes and the groups were miles apart from each other; the infection had to come from some non-human source. We looked for this in vain, in the armadillos, pacas, wild rodents, opossums and dogs, but the visit of 7 weeks was not long enough to solve the problem. Lainson and Strangeways-Dixon (1963) followed and showed that 3 arboreal rodents are the reservoirs (40 per cent *Ototylomys phyllotis*, 12 per cent *Nyctomys sumichrasti* and 10 per cent *Heteromys desmarestianus*). Further observations on the zoonotic aspects of this problem were made by the mammalogist Disney (1968) in subsequent years. All these workers noticed that the animals have leish-manial sores on their tails and that at night they descend from the trees to the ground where various species of *Phlebotomus* feed on them. The infection is thereby acquired by the insect which, 9 or 10 days later, may be passed on to a chicle worker sleeping in a hammock in the vicinity. A similar pattern of infection occurs in many other places in Central and South America, e.g. Guimarães (1966) incriminated another rodent, *Orizomys*, at Belém in the Amazonian forest. A single species of *Phlebotomus*— *Lutzomyia flaviscutellata*—has been shown to transmit the infection from Mexico (Biagi *et al.*, 1966) to Brazil.

The cutaneous infection over much of Latin America is sporadic, but it may become epidemic during population movements. Thus, 5 years ago, some Indians in the Mato Grosso were driven by another tribe to the Xingu River (de Carneri *et al.*, 1963), where Lainson and Shaw (1969) later found a high incidence of infection in various rodents, and a small epidemic broke out among the Indians on the borders of this river. Other Indians, who normally live in the high mountains around Lake Titicaca, sometimes descend at times of famine into the Amazonian regions in search of food—they obtain this but also contract leishmaniasis. In a revolution in Paraguay, the people took to the woods, and on their return to the cities, when the fighting was over, many of the people found that they were suffering from the disease. But the major factor in the creation of epidemics is the extensive deforestation of the country such as is seen in Brazil. The animal reservoir disappears with the alteration in the landscape, and other species of *Phlebotomus* transmit the infection between man and man. The disease in the terrible form of mucocutaneous leishmaniasis (espundia) becomes rife as the forest takes its revenge (Plate II, Fig. 9). The mild chiclero's ulcer of zoonotic origin is replaced by the sometimes fatal, purely human disease.

Great variations in the clinical and epidemiological patterns are thus found in the New World. Other types of interest are 'uta' which occurs in Peru and Chile at altitudes of up to 2800 m in the barren Andean cordilleras, where the dog is an important reservoir, 'pian-bois' in the Guyanas, and a peculiar form of leishmaniasis in Panama which has been much studied by workers at the Gorgas Institute (1966, 1967). Hertig *et al.* (1958) first showed that the organism could be cultured from the blood of 10 per cent of the spiny mice (*Proechimys semispinosus*) and that experimentally it could infect man, but not hamsters. In subsequent years, the infection could no longer be found in these animals, although man continued to contract the disease after visiting the forest; eventually the presence of *Leishmania* was revealed in various other wild animals including the kinkajou, porcupine and opossum, as well as in species of *Phlebotomus*.

The role of the animal reservoir is particularly puzzling in the New World and the fallacies which may arise in their interpretation have been discussed by the writer in detail (Garnham, 1965).

In contrast to cutaneous leishmaniasis, which is undoubtedly of indigenous origin, the visceral disease appears to have been imported into the New World after the Spanish conquest, when infected people and dogs crossed the ocean from Europe. The disease was scarcely recognized until 1934, when the introduction by Penna (1934) of a viscerotomy service for the diagnosis of fatal cases of yellow fever revealed its presence in Brazil. The brutal instrument, the viscerotome, is plunged through the abdominal wall of the cadaver into the liver; a piece of the organ is then removed, placed in formalin and subsequently examined for the characteristic lesions of yellow fever. Instead of yellow fever, *Leishmania chagasi* (= *donovani*) was sometimes found and the extension of this rather macabre service to other countries in Latin America (e.g. Colombia) revealed a widespread distribution of kala-azar. The actual incidence is not usually high, but cases have now been found from Mexico to Paraguay.

The landscape epidemiology is unlike that of the cutaneous disease; instead of forest, the background is rolling countryside at a low altitude and at the junction of hill and plain. The disease is more domestic and is seen mostly in small towns or settlements. Children are the chief sufferers and the work of Deane (1956) and Alencar (1958) in Brazil (Ceará) clearly demonstrated that kala-azar is a zoonosis, the former worker found a 12 per cent and the latter a 4 per cent infection rate in the fox, *Lycalopex* (= *Dusicyon*) *vetulus*. Dermal infection is intense in these animals, particularly around the muzzle, which is avidly attacked by the local vector (*Lu. longipalpis*). The fox often comes close to human dwellings to raid chicken runs; the *Lutzomyia* get infected, and the organism is transmitted to domestic dogs or, rarely, to cats or to man. The severity of the disease in the fox indicates that this animal is not a true indigenous reservoir, and the zoonotic picture is therefore not as transparent as was at first thought. The fox is uncommon in the Amazon Valley where sporadic cases of kala-azar were first noted 30 years ago by Evandro Chagas, and where *Dusicyon* was found infected by Lainson *et al.* (1969). With the involvement of dogs and foxes the infection becomes at times epizootic and epidemic, and the Brazilian investigators found that rates as high as 27 per cent occurred in dogs in certain foci.

Thus, although visceral leishmaniasis has a pronounced zoonotic aspect in South America, the zoonosis is likely to be an example of the process in reverse (see p. 10) as a result of the arrival of infected people into the New World.

African Trypanosomiasis

The polymorphic trypanosomes of African game animals provide a vivid example of the progress and eventual extinction of medical and veterinary zoonoses.

The original trypanosome, *T. brucei*, is transmitted from one antelope to another by the bite of tsetse flies of the savannah and the animals suffer little, if at all, from the infection which follows. This widespread belief was challenged recently at a symposium on diseases in free-living wild animals by Baker (1969), who suggested that little is really known about the effect on young animals in nature where perhaps many die from the disease. If man strays into the African bush, he may be bitten by infected flies; sooner or later, the trypanosome becomes adapted to human blood and an infection due to the so-called *T. rhodesiense* ensues, or strains, capable of infecting man, may have always accompanied the original *T. brucei*. Blair *et al.* (1968) describe sporadic cases in Rhodesia, occurring in hunters, fishermen and honey-gatherers, in which the disease is mild and terminates in premunition. Such 'healthy carriers' return to their villages where they provide a ready source of infection; the secondary cases which follow are liable to be more severe—they are the result of interhuman transmission.

It had long been suspected that antelopes provided a reservoir for the human type, because hunters or tourists came back from game expeditions in uninhabited country not only with trophies or photographs, but with sleeping sickness. The theory was only proved when Heisch *et al.* (1958) demonstrated that a trypanosome in a bush buck was infective to man.

Sporadic cases of a zoonotic origin are probably quite frequent in South-east Africa, but further north Ormerod (1963) points out how rhodesiense sleeping sickness becomes epidemic, the animal no longer remains a factor in the situation and interhuman transmission, still by the bite of tsetse flies occurs. In the severe epidemics which swept the Busoga Province of Uganda in 1940, the process had gone a step further, for the tsetse fly had largely disappeared and transmission, at least in some parts, was occurring mechanically by the transfer of infected blood on the proboscis of other biting flies, like *Stomoxys* or tabanids.

Further adaptation to man took place when peridomestic instead of savannah species of *Glossina* became involved; the remote *G. morsitans* and *G. swynnertoni* were replaced by *G. fuscipes* which breeds in riverine forest where human activities are numerous. Probably a slight change in the antigenic structure of the trypanosome occurred and the organism became the so-called *T. gambiense*, responsible for the more chronic form of sleeping sickness. There is indirect evidence that some domestic animals, such as pigs or goats, may acquire this form of the disease, but there is no longer any real animal reservoir (see p. 10). Possibly the mild cases of gambiense sleeping sickness, described by van Hoof (1947) in the Congo, Duggan (1962) from Nigeria and Harding and Hutchinson (1948) from Sierra Leone originated from such a source.

Eventually, though rarely, the trypanosome may pass congenitally from mother to child; details of eleven cases are summarized from the literature by Plainfosse et al. (1966) who describe a further instance in a child, nearly 3 years old, who developed neurological symptoms at 14 months, but remained undiagnosed until she was brought to hospital in a coma at the age of two years, when numerous trypanosomes were found in the blood.

The earlier evolution of these mammalian trypanosomes has been discussed by Baker (1963) and Hoare (1967). The latter points out the gradual improvements which have arisen in the mode of transmission; at first, in the monogenetic insect flagellates, the infective form is the cyst—eliminated in the faeces on to the inhospitable ground where it may, or more probably may not, be ingested by a new host; next, in the 'stercorarian' trypanosomes like *T. cruzi* which live in the blood of vertebrates, transmission is effected by biting insects, the infective form is again passed in the faeces and gets accidentally transferred to the mucous membrane of the new host; finally and best, in the 'salivarian' trypanosomes which have just been discussed, the organisms develop in the tsetse fly, reach its salivary glands and when the insect next bites, the infection inevitably enters the new host.

The zoonotic course of *T. brucei* is not entirely medical, for it may diverge into the veterinary field and again undergo a series of transformations.

Glossina morsitans transmits *T. brucei* of antelope origin to various domestic animals, including camels, cattle and horses, which develop a fatal form of the disease known as 'nagana' in Southern Africa. At the northern extremity of the tsetse belt, Hoare (1957) suggested that some domestic animals were taken outside it, and the infection was then passed from one animal to another mechanically, by the interrupted feeding of biting flies such as *Stomoxys*. On the Saharan border of Nigeria, it is

said by Godfrey and Killick-Kendrick (1960) that camel cara-
vans penetrate into the tsetse belt and later return northwards,
infected with *T. brucei* which is eventually transformed by
mechanical passage to *T. evansi*. Such trypanosomes lose their
polymorphic morphology, just as they do in syringe-passaged
strains in the laboratory. The disease is known as 'surra' in
camels, elephants and horses. Infections in horses may no longer
be transmitted by biting flies, but by venereal contact and the
disease 'dourine' is then the result.

The infection finally escaped from the continent of Africa and
acquired a cosmopolitan range. The Spanish conquistadores
imported horses from North Africa into the New World (see
Hoare, 1950); the horses took with them their African trypanosomes and the disease continued to spread by means of biting
flies in South America. At this point, a further morphological
change took place; the trypanosome lost its kinetoplast and then
became known as *T. equinum*, responsible for the condition *mal
de caderas*. In this continent transmission of infection took another
turn—the vampire bat took over the role of the horse fly, and,
moreover, itself became a reservoir of infection. With its razor-
edged incisors, the vampire makes a shallow wound on the
horse's shoulder. The bite is unnoticed, blood flows freely and
the vampire laps it up, together with the trypanosomes. Later
the bat bites another horse and a new victim of the infection is
claimed. In the words of Scott:

> For like the bat of Indian brakes
> Her pinions fan the wound she makes,
> And soothing thus the dreamer's pains,
> She drinks the life-blood from the veins.

Scott would have added another lurid verse if he had known
that the vampire may transmit to its victim both trypanosomiasis *and* rabies. Scott was ignorant, but the Indians of Darien
early in the sixteenth century were aware of the association and
used to cauterize the wound made by the bat as a prophylactic
measure (Marinkelle, 1968).

The pattern of sleeping sickness in Africa has changed in
recent years as the result of increased communications, and

these further developments are discussed below in Chapter 5 with more specific reference to their public health aspects.

New World Trypanosomiasis

Chagas' disease is caused by infection with *Trypanosoma cruzi* and transmission is effected by the bite of various species of reduviid bugs, of which the three most important are *Triatoma infestans*, *Panstrongylus megistus* and *Rhodnius prolixus*. The infection extends from Maryland to the Argentine but is only serious in the more tropical lands. In its northern range, man is never affected and the infection is maintained in wild animals; in the more primitive conditions further south, contact between man, reservoir and vector is closer and millions of human infections occur. In the latter circumstances, the condition is acquired in early childhood and is often confined to mild fever, lymphadenopathy and a primary lesion (chagoma) on an exposed part of the body or an unilateral swelling of the eyelid (Romaña's sign). Then the trypanosome retreats to an internal organ such as the heart or alimentary tract, where it multiplies in 'nests' in the muscle and eventually, 40 years or so later, produces heart failure or other lethal conditions. Only in recent years has the pathology of this mysterious condition been elucidated; Köberle (1968) finally demonstrated that a neuro-toxic substance is liberated when the parasites degenerate (*not* when the pseudocyst ruptures) which attacks the autonomic ganglion cells, ruining the nervous control of peristalsis and destroying the nerve cells of the heart. Köberle emphasizes that these late sequelae represent the final consequence of the disease proper—they are not directly of parasitic origin, but neuropathies arising from a reduction in the number of ganglion cells.

The Chagasian landscape is characteristic. It is rural and unforested and contains poor dwellings made of mud and wattle or other wooden planks. In the cracks of the walls, the reduviid bugs breed in enormous numbers, and at night, emerge to bite the sleeping inhabitants. The endemic area lies below 1000 metres, has a fairly low rainfall, and is the opposite extreme to the tropical rain-forests of the Amazon and Orinoco basins, the home of leishmaniasis.

In Maryland, the infection is limited to the racoon and bug;

MANKATO STATE COLLEGE LIBRARY
Mankato, Minnesota

further south in Texas, the woodrats are the reservoir and only very rarely has man acquired the disease. In British Honduras, Petana (1969) demonstrated the existence of *T. cruzi* in *Triatoma dimidiata*, and Coura (1967) in man by serological reactions and electrocardiographic abnormalities. The opossum was incriminated as a reservoir by Andrew Robertson (1930) in Spanish Honduras, and bats have been found infected in Panama.

The great home of the disease is South America, and in Brazil the major wild animal reservoirs are armadillos and opossums; also rodents and perhaps monkeys. In Bolivia, the cavy plays a similar role.

These animals form the sylvatic reservoir; reduviid bugs breed in their burrows and transmit the infection in a continuous cycle. In rural settlements, infected bugs fly to the huts, which they colonize and transmit the trypanosome to the inhabitants, including the domestic animals. In one such endemic area in the State of São Paulo, not far from the Rio Grande, the writer saw typical dwellings in which the incidence of the infection was 19 per cent in the dogs, 16 per cent in the cats, and nearly 100 per cent in the triatomid bugs (de Freitas, 1955). Probably the human population is as important a reservoir as the domestic animal in these circumstances, and the zoonotic connection becomes unnecessary. The armadillo disappears from the scene early, then the dogs and cats, and eventually the invertebrate vector; congenital transmission is not at all uncommon in places of high endemicity (Gavaller, 1951; Rubio *et al.*, 1961).

Once the housing improves, the infection diminishes, and such a picture is well delineated in small towns; on their rural outskirts the rate is high but it rapidly sinks towards the centre where there are better buildings. The bug is easily exterminated by the use of gammexane, and the more intelligent inhabitants make use of this insecticide to get rid of the pest.

BACTERIAL INFECTIONS

Pavlovsky (1966) took the following bacterial diseases to illustrate his theory of natural focality: plague, tularaemia, brucellosis, anthrax and listerellosis, and of these, plague is selected here as providing the best example of the complete process; the other conditions usually come to a premature end.

Plague

The first microbial disease to be recognized as a zoonosis was probably plague; yet in split of the violence of the epidemics for which it is responsible and the antiquity of the records, the link with animals is often inconspicuous and ill-recognized. In mediaeval times, the population was so decimated by the disease, that the connection with rats passed entirely unnoticed, let alone the true origin of the infection in the wild rodents. Daniel Defoe (1960) in his vivid and detailed account of the great plague which swept London and later most of England soon after the Restoration was apparently quite unaware of any mortality in the rats.

Even the keen observers in those illustrious centres of emerging science of Bologna, Pavia and Padua, failed to find the key to the problem, though they were clever enough sometimes to halt the passage of the infection. The author wrote these lines at Bellagio in a look-out, built in 1629, at the request of the Tribunal of Health, Milan by the Duke Sfondrati; guards armed with guns and small cannon were posted there to prevent strangers from introducing plague into the town (Miglio, 1959 and 1966 and Adami, 1927). The disease had reached the district with German troops from over the Alps (in another invasion of Italy) and the strict isolation of Bellagio protected this town, alone of all others on the Lago di Como, from plague, for not a single case occurred here. Across the lake at Bellano, lived the famous doctor and philosopher, Sigismondo Boldoni; the epidemic struck this village and Boldoni himself died of the plague. This was the village where 268 years later, Grassi wrote to his daughter to tell her that he had found the mosquito responsible for malaria (see p. 163).

Plague continued its course southwards and reached Lecco at the extremity of the lake. Here still stands the house (Il Caleotto) of Manzoni who describes in his famous novel, 'I Promessi Sposi', this actual epidemic (again with no mention of rats) which devastated Milan, killing almost 200 000 people. By a strange coincidence, Manzoni sold Il Caleotto in 1818 to a Signore Scola, the great-great-grandfather of Grassi's niece, Signora Grassi-Scola who lives there today and who maintains on its ground floor the relics of Manzoni and on the upper, those of her uncle.

In China, as early as the eighteenth century epidemics of plague were associated in the minds of the inhabitants with an unusual mortality in house rats (see Hoeppli, 1959). And it was in China that the true solution was found early in the present century: plague is a zoonosis with all the characters that have been described above. Its original sylvatic form usually disappears early, as in the major epidemics of history; the epizootics in domestic rats, which must accompany or precede the latter may escape attention, and sometimes it is difficult to understand the mechanism of geographical spread. The habits of the people, the rodents, the fleas; the climatic conditions; and the transport of goods must all be studied (see, for East Indian conditions, Dinger, 1964 and the original observations of his predecessor, Swellengrebel, 1913).

The infection primarily occurs in a variety of wild rodents (Garnham, 1949) such as the tarabagan in Mongolia, *Tatera* in Africa and ground squirrels in California. Hundreds of species harbour the infection, many do not suffer much from the disease and, like the gerbils resistant to *Leishmania* in Iran, or the bush-buck infected with trypanosomes in Kenya, constitute ideal reservoirs of infection.

Man acquires the infection only by accident if he intrudes into the area either whilst hunting, collecting fire-wood or for some such purpose. Normally transition to man occurs in the following way: semi-domestic rodents such as *Arvicanthis* or *Praomys coucha* act as liaison carriers, they stray into the veldt, pick up plague-infected fleas, develop the infection and bring it to the vicinity of human dwellings. The fleas drop off the dying animals and attach themselves to the domestic rodents. The latter die quickly from the infection and their infected fleas jump on to man and give him bubonic plague. Next, the human flea (*Pulex irritans*)—according to Blanc (1956)—transmits the organism from man to man, and the zoonosis disappears.

Finally under conditions of overcrowding, or in other adverse circumstances, the disease passes directly from man to man as an airborne infection, giving rise to pneumonic plague.

The late Georges Blanc of Casablanca was the prototype of the enlightened doctor (Bernard Rieux) of Camus' great novel or epidemiological treatise *La Peste*. From the apparently insignificant discovery of a dead rat, Rieux watched with horror the refusal at first of the government of Oran to acknowledge the nature of the epidemic. The description is so vivid that it surpasses even Manzoni's account of the plague in Milan. Camus wrote his book in the early 1940s, just before DDT and streptomycin became available, since when such disasters have become relics of the past.

The domestic rat is very susceptible to *Pasteurella* (= *Yersinia*) *pestis* and the mortality is so high that the epizootic in these rodents eventually burns itself out. The bubonic plague acquired from the veldt is usually rather less severe than the domestic form, but when pneumonic plague supervenes, with no intervention of rodent or flea, the case assumes a more virulent aspect and is practically always fatal. Once again occurs the increase in virulence as humanization of the strain progresses and the zoonosis vanishes.

Davis *et al.* (1968) have drawn attention to the curious silent intervals between epizootics among the wild and domestic rodents, and it seems that the infection is dormant at this time, only to be lit up by an unknown stimulus, such as an arbovirus of group B. These authors are inclined to accept the view that three varieties of the organism exist (*antiqua*, *mediaevalis* and *orientalis*) which stem from the historical outbreaks, and possibly some of the epidemiological puzzles can be explained on this basis.

This is the normal course of events in the development of plague; in modern life where man has upset the slow tempo of natural phenomena, a strange reversal of the picture is seen. Plague probably used not to exist in South Africa or the United States, but ships from the tropics brought plague-infected rats to Durban and San Francisco respectively. They escaped on to the quays and gave rise to epizootics in the local rats which in their turn were carried by train in merchandise to the hinterland.

Once there the plague spread into the wild rodents and plague travelled further and further afield as a sylvatic infection reverting to its original type but in a new locality.

SPIROCHAETAL DISEASES

Relapsing Fever

Relapsing fever spirochaetes are of two main types, one of which is tick-borne and the other louse-borne. The former occurs in many varieties which are peculiar to certain geographical regions, and these probably represent the original organisms which were parasites of soft ticks. The tick may be regarded as the more important reservoir host of the infection, which persists in these arthropods for long periods and may carry over into subsequent generations by transovarial passage.

The natural focus is comprised by caves or burrows in remote places, where wild rodents of various kinds live in association with different species of *Ornithodorus* ticks. The ticks are infected with spirochaetes (*Borrelia* spp.) which they transmit when they feed on these animals. Sometimes man intrudes into this environment and becomes infected with the organism. Thus, the late Saul Adler once slept in a cave in the hills above the Plain of Sharon in Israel and was bitten by *Ornithodorus papillipes*; on his return to Jerusalem some days later, he developed relapsing fever (Sheba, 1968). Shortly after this episode, General Morrison (1937) described a small epidemic acquired in the same place by soldiers who had been benighted in an open cave or dug-out. Such infestations are to be found all along the North African littoral, where *O. erraticus* transmits *B. hispanica*, *B. crocidurae* and *B. merionesi* to the shrews, merions, gerbils, wild rabbits, and occasionally to man.

Other wild rodent spirochaetes (e.g. *B. dipodilli* of the pigmy gerbil) and different species of *Ornithodorus* (e.g. *O. graingeri*) are found in tropical Africa. Heisch (1950) showed that *B. dipodilli* is infective to man and concluded that the well known human spirochaete—*B. duttoni*—may have evolved from such a source. The organism becomes transmitted by the domestic tick, *O. moubata*, which is an excellent host for the parasite, and is passed directly to man without the intervention of another mammalian host. In other words the zoonosis has been replaced by inter-human transmission. Even in the desert areas of northern Kenya and Uganda and in Somalia, where the human population is sparse and the possible vector (*O. savignyi*) is non-domestic, relapsing fever does not revert to the zoonotic state.

The next stage of the process occurs when the louse starts to transmit the organism and louse-borne relapsing fever breaks out in epidemic or pandemic form. Such a transformation seems to take place in unsettled times during famines and large-scale migrations. A severe pandemic originated towards the end of the Second World War in the Fezzan in Libya where, as in the Ethiopian highlands, louse-borne relapsing fever is endemic. The infection spread along caravan routes and by military movements throughout North Africa, into upper Egypt (Kamal *et al.*, 1947), the Middle East, Iran and Arabia. It finally reached East Africa by dhow across the Indian Ocean and produced an epidemic with many deaths in the Coast Province of Kenya. The application of DDT dust by a well organized medical service quickly halted the further progress of the disease (Garnham *et al.*, 1947).

Once again we observe the enhancement of virulence of a parasite as it passes from the mild form acquired from animals to the lethal type produced by direct inter-human transmission via man's own ectoparasite the louse. As in other examples, insect transmission may be finally eliminated with the passage of the organism through the placenta. The writer observed a congenital case in Kisumu, Kenya in 1936; the mother became ill with relapsing fever the day after she gave birth to a child (with retention of the placenta); the child showed parasites in the blood 9 days later, and it was difficult to decide if infection had taken place through the placenta or during parturition.

The parasite presumably changes its antigenic composition during this hypothetical evolution: its name changes from *B. crocidurae* to *B. duttoni* and finally to *B. recurrentis* in the louse-borne form. Heisch and Garnham (1949) demonstrated by experiments in the laboratory that *B. duttoni* could be transmitted by lice instead of ticks. Some factor seems to operate in nature, however, which prevents the ready conversion of tick-borne to louse-borne relapsing fever.

These speculations on the evolution and the different varieties of relapsing fever are not intended to cover the complete epidemiology of the disease. The American infections have been omitted from consideration; moreover the suggested course of the tropical African *B. duttoni* from a crocidurae type of organism is unlikely to be the whole story. Geigy (1968) like Heisch, considers the possibility that *B. duttoni* may have a phylogenetic connection with one or more spirochaetes found in different species of *Ornithodorus* which feed on rodents and other mammalian hosts, as found in South Africa; this represents the southern parallel of the North African hypothesis. But Geigy is more inclined to favour the idea that *B. duttoni* has evolved from the human *B. recurrentis*; this is quite contrary to the general theory concerning the zoonosis, which assumes that man comes last rather than first in the chain. It is much easier to believe that the European and Middle East epidemics of louse-borne relapsing fever originated from the Hispano-African tick-borne forms, rather than vice versa.

These theories place the chief emphasis on rodents as the original mammalian host of human infection. It is possible that the lower primates may have played a part, as species of *Borrelia* have occasionally been found in these animals, e.g. *Borrelia harveyi* (Garnham, 1947) in *Cercopithecus mitis* in the forests on the Mau Escarpment above the Rift Valley in East Africa, and close to the famous sites of origin of primitive man.

RICKETTSIAL INFECTIONS

The natural history of rickettsiosis is similar in many ways to that of relapsing fever in that the condition is originally of a hetero-geneous cosmopolitan nature in ticks and mites, later affecting rodents and eventually man; mutations probably occurred with the final production of the louse-borne form of epidemic typhus. The allied infection, Q fever, was first thought to be rickettsial in nature, but the causative organism has now been placed in a new genus, as *Coxiella burneti*; its zoonotic course resembles that of the true rickettsioses in that hard ticks and rodents are the feral hosts, but differs in that domestic animals such as cows, sheep and goats contract the infection and pass it on to man

PLATE III

Fig. 11. The Kuja River at 1700 m: near the Country of the Blind (focus of onchocerciasis).

Fig. 12. Vegetation around native hut in Central Nyanza: a new focus of Rhodesian sleeping sickness.

PLATE IV

FIG. 13. Dawn on the Sesse Islands: the original focus of Gambian sleeping sickness in Uganda.

FIG. 14. A peaceful scene in Galway (Lough Corrib): cattle reservoir of human piroplasmosis and where, in his old age, Manson used to fish.

either via the milk or more often by inhalation of the dried organism from organs or excreta of infected animals (Evenchik, 1964).

Audy (1968) suggests that the first link in the typhus chain was the mite which harboured *Rickettsia tsutsugamushi* and passed it on to rodents. According to him, the tick-borne infections came second. Both arthropods are ectoparasitic on wild rodents in the Far East, Africa and the New World and the animals at some time become infected and harbour the rickettsiae. Man may become infected directly by being bitten by the arthropods whilst passing through infested bush; the foci of infection are limited, in the case of scrub typhus, to very small areas—the so-called mite islands—as the result of special ecological factors.

In the case of tick typhus, infection is acquired either by bites of ticks in the veldt, or the tick is brought into human dwellings on dogs which act as liaison carriers and the tick then bites the human occupants.

Only ixodid ticks are involved, and these convey the human diseases, Rocky Mountain spotted fever (due to *R. rickettsi*), fièvre boutonneuse of the Mediterranean littoral, Kenya tick typhus (due to *R. conori*) and other closely related, if not identical infections, from South America to Siberia.

The feral aspects of Kenya tick typhus were studied in considerable detail by Heisch *et al.* (1962), who showed that the vector ticks were *Haemaphysalis leachi*, *Rhipicephalus simus* and *Amblyomma variegatum*, which are found both on dogs and in rodent burrows. *Rickettsia conori* and *C. burneti* were isolated from various wild rodents and from *Rattus rattus*.

In the black rat, a change in the antigenic properties of the organism apparently takes place, it becomes transformed into *R. mooseri* and is transmitted by fleas instead of ticks. Murine typhus is the product, a much more domestic and urban form of the disease than the previous varieties. Another form of typhus with a reservoir in rats (and mice) is rickettsial pox, caused by *R. akari* and transmitted by mites. This is again a disease of cities in North America and Russia.

Later still, another transformation takes place and has actually been witnessed in Mexico (see Burnet, 1966): under crowded conditions, man's own ectoparasites, the body lice, acquire the infection and fulminating epidemics of louse-borne typhus, due to *R. prowazeki*, are the next stage of the process. Here the zoonosis has disappeared, and finally the vector itself may vanish; the organisms are fairly resistant and they manage to survive in the dust on the floors of houses—the dust is subsequently inhaled and man contracts the infection in this way.

R. prowazeki is usually regarded as a purely human infection, transmitted by lice. However in 1967, Ruth Reiss-Gutfreund reported the isolation of a strain of this parasite from cattle in the Ethiopian highlands and from their ticks (*Hyalomma dromedarii*). Although other workers have failed to repeat this observation, the existence of a bovine reservoir certainly offers an explanation for the well-known persistence of this infection in Ethiopia.

VIRUS INFECTIONS

Virus diseases of many kinds have a zoonotic interest; several have been mentioned already, such as rabies which is always of such an origin and influenza which probably stemmed from animal infections (e.g. in chickens, ducks, pigs and horses) in past eras. But the arboviruses have the most significance in parasitology because their pattern is typical of the whole group, e.g. their occurrence in natural foci, the ever-changing pattern, the frequency of inapparent infections when acquired from a feral source, and of course the zoonotic element.

The epidemiology of a few of these conditions has been most extensively studied, but in most the true relations between man, a succession of wild animal and domestic hosts, and a bewildering array of vectors remain largely unknown, and we are faced with the probability of hidden zoonoses of viral origin and undergoing dangerous mutations as an ever-present threat to man's existence. It is always alarming when a 'new' virus suddenly attacks man, like B virus and vervet virus (Martini, Simpson, and Zlotnik, 1969) of monkeys, or when an epizootic of unknown aetiology affects the animals and spreads as a severe affliction to man. Typical examples of the latter zoonoses are Kyasanur forest disease (Work *et al.*, 1957) with the monkeys dying in the forests of Mysore and a prostrating sometimes fatal disease of man, and Rift Valley fever (Daubney *et al.*, 1931) which causes epizootic hepatitis in sheep in Kenya and abortions in cattle in South Africa and a highly infectious disease in man; among the original workers, the writer was about the only person to escape the malady. In both these examples, the true zoonotic origin has still not been determined, but it is probably the wild rodent.

Other arboviruses also have a succession of animal hosts before they reach man. Russian spring-summer encephalitis is thought to originate in wild rodents, and pass to birds and man via ixodid ticks in the *taiga* of Siberia; the so-called 'equine' encephalitides probably start in birds such as pigeons and pheasants and reach horses and finally man by the bite of culicine mosquitoes. Japanese B-encephalitis is another zoonosis of similar type, running a course through wild birds, swine and other animals to man; the widespread West Nile fever has its chief reservoir according to Taylor *et al.* (1956) in birds (particularly crows and sparrows), and a variety of animals including monkeys, horse, sheep and man become similarly infected, probably as 'dead end' cases, as in most of the above diseases.

Yellow Fever

The natural history of yellow fever is much better known than that of the infections briefly cited above, and it provides an excellent example of the progression of a zoonosis on the widest scale.

The virus of yellow fever may originally have been localized in mites, which perhaps transmitted the infection to the galago or bush baby. The sera of these animals have often been found to contain antibodies against yellow fever by the mouse-protection test; Lumsden (1956) and Haddow and Ellice (1964) found an immunity rate of 23 per cent in Zambia and Malawi, and showed that such animals when challenged with yellow fever virus were immune, though bush babies without antibodies became heavily infected. Lumsden (1953), after years of patient work, failed to incriminate the mite as a vector in laboratory experiments, though the virus survived for 4 days in this arthropod.

If not the mite, canopy-dwelling or acrodendrophilic mosquitoes presumably become infected from the galago and pass the
virus to the monkey population (though the exact order in
which these animals acquire their infections is really unknown).
The *Colobus* monkey and the red tail in particular inhabit this
upper stratum of the forest and these species show a high
immunity rate (53 per cent–Haddow *et al.*, 1951) as a consequence of attack by the acrodendrophilic mosquito, *Aedes
africanus*. The mangabeys and blue monkeys become infected to
a lesser extent (44 per cent) in the midzone by the same species
of mosquito and the infection is eventually passed to *Cercopithecus
aethiops*, a monkey which spends much of its time on the ground,
though it sleeps at night in low forks of the trees. These pre-
dominantly terrestrial monkeys have the lowest immunity rates
(35 per cent). *C. aethiops* raids native plantations and gets
bitten by a mosquito, *Aedes simpsoni*, which breeds in water in
the axils of banana trees. These mosquitoes subsequently bite
the local people who develop yellow fever, often of a fairly mild
nature as recovery in the indigenous African is not uncommon
(Haddow, 1965).

A person during the incubation period of jungle yellow fever
may then visit a neighbouring town, infested perhaps with the
domestic mosquito, *Aedes aegypti*. The virus begins to circulate in
the blood of this man, the mosquitoes bite him, and an epidemic
of urban yellow fever follows. The zoonosis has disappeared and
the sylvatic vector has been replaced by a domestic one. Finally,
the mosquito vector may drop out of the picture and transmission may rarely take place directly from man to man, either
transplacentally to the foetus (Sicé and Rodallec, 1940) or by
contact with infective material, particularly blood as in taking
specimens for clinical pathology or organs as in conducting
autopsies (see Low and Fairley, 1931 etc.). At one time, it was
firmly believed that yellow fever could be readily contracted by
the handling of soiled linen, clothing, etc., and in the first edition
of *Tropical Diseases*, Manson (1899) states that the 'destruction
of all fomites is imperative'. A year or two later, the Walter
Reed Commission (see Strode, 1951) entirely disproved this idea
by showing that yellow fever is transmitted by the bite of
infected mosquitoes and is not conveyed by fomites. These

famous experiments required the use of human volunteers and some members of the Commission died from the self inflicted disease. These transformations are seen to involve a gradual increase of virulence of the parasite and there is always a heavy mortality in urban epidemics of the disease.

Yellow fever is thought to have been carried in *Aedes*-infested slavers to the New World, perhaps as early as the nineteenth century, when a zoonosis in reverse became established. The monkeys of South and Central America acquired the disease from the imported source, and because these species had had no experience of the virus, unlike the African animals, they died in thousands as the epizootics spread through the forests of Latin America. It is a sombre experience to wander in a forest through which yellow fever has recently passed; it is quite silent instead of ringing with the long drawn-out howl of the majestic *Alouatta* in the tree-tops, or the chattering and barking of the capuchins and spider-monkeys lower down—sounds which the Brazilian composer, Villa-Lobos, has incorporated in his great symphonic poem *Amazonas*.

Yellow fever in tropical America is transmitted by the bite of acrodendrophilic mosquitoes, *Haemogogus spegganzinii* and *H. capricorni* and sabethines of gorgeous metallic lustre. Forest workers acquire the disease and on their return to civilization may pass it on to *Aedes aegypti*. An urban epidemic used to be the inevitable result, until the vigorous anti-amaryl measures of the Rockefeller Foundation largely removed this menace from the New World.

The elucidation of the epidemiology of yellow fever has occupied most of the present century, first its transmission by mosquitoes, then observations on the virus, next the discovery of jungle yellow fever and the role of monkeys and finally the extraordinarily complicated mosquito cycles in the forests of Africa and Latin America (see Soper, 1970, ref. in Addenda). This research has been dependent upon a close study of the natural foci of the infection, including the habits of the monkeys and of the sylvatic mosquitoes. The clue to the role of mosquitoes was found by Bugher *et al.* (1944) when these North American and Colombian investigators were walking in the forest near Villavicencio; a tall tree was felled by wood-cutters, it crashed to the ground and brought from its topmost branches myriads of *Haemogogus* which at once proceeded to bite everyone in the vicinity. The implication was obvious. The introduction of the tree platform for observations on acrodendrophilic mosquitoes has been an essential element for success, and Haddow's (1965, 1968) work in tropical Africa has been of the greatest significance. Apart from studies on innumerable tree platforms, his 24 hours' catches at different levels of a steel tower, 36 m high, in the Zika Forest of Uganda illustrated the behaviour of mosquitoes and other biting insects in the minutest detail.

CONTROL OF ZOONOSES

Observations of this kind on the different factors in a natural focus of any disease are necessary for successful control. A correct assessment of the zoonotic situation is essential, for the existence of a primitive reservoir might remain entirely unsuspected, and eradication measures directed only to the causative agent and its vector could be quite futile. Direct attack on the animal reservoir is seldom feasible; it would be impracticable to remove the monkeys or wild rodents from South America in order to halt yellow fever or leishmaniasis, and it is immoral, though still done in some countries, e.g. Rhodesia and Uganda, to kill the antelopes in order to prevent trypanosomiasis. These beautiful animals are being slaughtered in the former country at the rate of nearly 10 000 a year (*Ann. Rept* 1968). On the other hand, rodent control against plague in cities is a long established practice, and fumigation of the gerbil burrows with chloropicrin in Turkmenia has been entirely successful in the eradication of the zoonotic form of oriental sore from that region (Latyshev and Kryukova, 1941). The simple answer to the problem is sometimes provided by vaccination of the population at risk (against yellow fever or leishmaniasis).

REFERENCES

ADAMI, V. (1927)* *Varenna e Monte di Varenna*, Milan.
ADLER, S. (1964) In *Advances in Parasitology*, vol. 2, 'Leishmaniasis', pp. 35–96, London: Academic Press.
ADLER, S. and ADLER, J. (1955) 'The agglutogenic properties of various stages of the Leishmanias', *Bull. Res. Counc. of Israel* **4**, 396–7.

ALENCAR, J. E. (1958) 'Aspectos clinicos do cala-azar Americano', *Proc. 6th Internat. Congr. trop. Med. Malaria* III, 718–46.

ANDRÉ, R., BRUMPT, L., DREYFUS, B., PASSELEC, Q. and JACOB, S. (1957) 'Leishmaniose cutanée, leishmaniose cutanéo-ganglionnaire et kala-azar transfusionnel', *Bull. Mem. Soc. Méd. Hopitaux Paris* 25, 854–61.

ANNUAL REPORT (1968) Branch of tsetse and trypanosomiasis control, Rhodesia, Salisbury: Government Printer.

AUDY, J. R. (1968) *Red Mites and Typhus* (Heath Clark Lectures, 1965) University of London, The Athlone Press.

BAKER, J. R. (1963) 'Speculations on the evolution of the family Trypanosomatidae Doflein, 1901', *Exp. Parasit.* 13, 219–33.

BAKER, J. R. (1969) 'Trypanosomes of wild mammals in the Serengeti' in *Symposium No. 24 Zoological Soc. London*, Academic Press.

BLAIR, D. M., SMITH, E. B. and GELFAND, M. (1968) 'Human trypanosomes in Rhodesia', *Central Afr. J. Med.* 14, suppl. no. 7, 1–12.

BLANC, G. (1955) 'Conservation dans la nature de virus pathogènes pour l'homme. Une observation marocaine', *Maroc. méd.* 34, 933–7.

BUGHER, J. C., BOSHELL, M. J., GARCIA, R. M. and MESA, O. E. (1944) 'Epidemiology of jungle yellow fever in eastern Colombia', *Am. J. Hyg.* 39, 16–51.

BURNET, M. (1966) *Natural History of Infectious Disease*, Cambridge: University Press.

CAHILL, K. M. (1964) 'Leishmaniasis in the Sudan Republic, XXI. Infection of American personnel', *Am. J. trop. Med. Hyg.* 13, 794–9.

CAMUS, A. (1947) *La Peste*, Paris: Flammarion.

DE CARNERI, I. (1903) 'Epidemia de leishmaniose tegumentar entre dos Indios Waura do Parque Nacional do Xingu', *Rev. Inst. Med. trop. São Paulo* 5, 271–2.

COURA, J. R. and PETANA, W. B. (1967) 'American trypanosomiasis in British-Honduras 11.—The prevalence of Chagas' disease in Cayo District', *Ann. trop. Med. Parasit.* 61, 244–50.

DAUBNEY, R., HUDSON, J. R. and GARNHAM, P. C. C. (1931) 'Enzootic hepatitis or Rift Valley Fever', *J. Path. Bact.* 34, 545–58.

DAVIS, D. H. S., HEISCH, R. B., McNEILL, D. and MEYER, K. F. (1968) 'Serological survey of plague in rodents and other small mammals in Kenya', *Trans. R. Soc. trop. Med. Hyg.* 62, 838–61.

DEANE, L. M. (1956) *Leishmaniose visceral no Brasil*, Servicio Nacional de Educacão Sanitaria, Rio de Janeiro.

DEFOE, D. (1960) *A Journal of the Plague Year*, London: Folio Soc.

DINGER, J. E. (1964) In *Zoonosis*, Ed. J. van der Hoeden, Amsterdam: Elsevier.

DISNEY, R. H. L. (1968) 'Observations on a zoonosis: Leishmaniasis in British Honduras', *J. appl. Ecol.* 5, 1–59.

DUGGAN, A. J. (1962) 'A survey of sleeping sickness in Northern Nigeria from the earliest times to the present day', *Trans. R. Soc. trop. Med. Hyg.* 56, 439.

EVENCHICK, Z. (1967) 'Q-fever' in *Zoonoses*, Ed. J. van der Hoeden, Amsterdam: Elsevier.

FREITAS, J. P. DE (1955) Personal communication.

GARNHAM, P. C. C. (1936) 'A case of "congenital" relapsing fever', *E. Afr. med. J.* **15**, 50–51.

GARNHAM, P. C. C. (1947) 'A new blood spirochaete in the grivet monkey, *Cercopithecus aethiops'*, *E. Afr. med. J.* **24**, 47–51.

GARNHAM, P. C. C. (1949) 'Distribution of wild-rodent plague', *Bull. Wld Hlth Org.* **2**, 271–8.

GARNHAM, P. C. C. (1965) 'The Leishmanias, with special reference to the role of animal reservoirs', *Am. Zool.* **5**, 141–51.

GARNHAM, P. C. C., DAVIES, C. W., HEISCH, R. B. and TIMMS, G. L. (1947) 'An epidemic of louse-borne relapsing fever in Kenya', *Trans. R. Soc. trop. Med. Hyg.* **41**, 141–70.

GARNHAM, P. C. C. and LEWIS, D. J. (1959) 'Parasites of British Honduras with special reference to Leishmaniasis', *Trans. R. Soc. trop. Med. Hyg.* **53**, 12–40.

GAVALLER, B. (1951) 'Enfermedad de Chagas congenità', *Bol. Matern. Concepcion Palacios* **4**, 59–64.

GODFREY, D. G. and KILLICK-KENDRICK, R. (1962) '*Trypanosoma evansi* of camels in Nigeria', *Ann. trop. Med. Parasit.* **56**, 14–19.

GORGAS MEMORIAL LABORATORY (1966) 'Leishmaniasis transmission-reservoir studies', *Annual Report* **37**, 11–15.

GUIMARÃES, F. N. (1966) '*Oryzomys goeldii*, a wild rat from Amazonia, as reservoir of *Leishmania brasiliensis'*, *Proc. 1st Int. Congr. Parasit.* Rome.

GUNDERS, A. E., FONER, A. and MONTILLO, B. (1968) 'Identification of *Leishmania* sp. isolated from rodents in Israel', *Nature* **219**, 85–86.

HADDOW, A. J. (1965) 'Yellow Fever in Central Uganda, 1964', *Trans. R. Soc. trop. Med. Hyg.* **59**, 436–58.

HADDOW, A. J., CASLEY, D. J. L., O'SULLIVAN, P., ARDOIN, P. M. L., SSENKUBUGE, Y. and KITAMA, A. (1968) 'Entomological studies from a high steel tower in Zika Forest, Uganda. Part II. The biting activity of mosquitoes above the forest canopy in the hour after sunset', *Trans. R. entomol. Soc.* **120**, 219–36.

HADDOW, A. J., DICK, G. W. A., LUMSDEN, W. H. B. and SMITHBURN, K. C. (1951) 'Monkeys in relation to the epidemiology of yellow fever in Uganda', *Trans. R. Soc. trop. Med. Hyg.* **45**, 189–224.

HADDOW, A. J. and ELLICE, J. N. (1964) 'Studies on bush-babies with special reference to the epidemiology of yellow fever', *Trans. R. Soc. trop. Med. Hyg.* **58**, 521–38.

HADDOW, A. J. and SSENKUBUGE, Y. (1965) 'Entomological studies from a high steel tower in Zika Forest, Uganda. Part I. The biting activity of mosquitoes and Tabanids as shown by 24 hour catches', *Trans. R. ent. Soc. Lond.* **117**, 215–43.

HARDING, R. D. and HUTCHINSON, M. P. (1948) 'Sleeping sickness of an unusual type in Sierra Leone and its attempted control', *Trans. R. Soc. trop. Med. Hyg.* **41**, 481–512.

HEISCH, R. B. (1950) 'On *Spirochaeta dipodilli* sp.nov., a parasite of pygmy gerbils (*Dipodillus* sp.)', *Ann. trop. Med. Parasit.* **44**, 260–72.

HEISCH, R. B. (1956) 'Zoonoses as a study in ecology', *Brit. med. J.* ii, 669–673.

HEISCH, R. B. (1963) 'Is there an animal reservoir of kala-azar in Kenya?', *E. Afr. med. J.* **40**, 359–63.

HEISCH, R. B. and CANNHAM, P. C. C. (1948) 'The transmission of *Spirochaeta duttoni* by *Pediculus humanus corporis*', *Parasitology* **38**, 247–52.

HEISCH, R. B., GRAINGER, W. E., HARVEY, A. E. C. and LISTER, G. (1962) 'Feral aspects of rickettsial infection in Kenya', *Trans. R. Soc. trop. Med. Hyg.* **56**, 271–86.

HEISCH, R. B., GRAINGER, W. E. and D'SOUZA, J. ST A. M. (1953) 'Results of a plague investigation in Kenya', *Trans. R. Soc. trop. Med. Hyg.* **47**, 503–21.

HEISCH, R. B., McMAHON, J. P. and MANSON-BAHR, P. E. C. (1958) 'The isolation of *Trypanosoma rhodesiense* from a bushbuck', *Br. med. J.* ii, 1203–4.

HERTIG, M., FAIRCHILD, G. B. and JOHNSON, C. M. (1958) 'Leishmaniasis transmission-reservoir project', *Rep. Gorgas Memorial Lab.* **30**, 7–11.

HINDLE, E. and THOMSON, J. G. (1928) 'Leishmania infantum in Chinese hamsters', *Proc. Roy. Soc. B.* **103**, 252–7.

HOARE, C. A. (1950) 'Akinetoplastic strains of *Trypanosoma evansi* and the status of allied trypanosomes in America', *Rev. Soc. med. Hist. nat.* **10**, 81–90.

HOARE, C. A. (1957) 'The spread of African trypanosomes beyond their natural range', *Z. Tropenmed.* **8**, 157–61.

HOARE, C. A. (1967) 'Evolutionary trends in mammalian trypanosomes' in *Advances in Parasitology*, vol. 5, London: Academic Press.

HOEPPLI, R. (1959) *Parasites and parasitic infections in early medicine and science*, Singapore: University of Malaya Press.

HOOF, L. VAN (1947) 'Observations on trypanosomiasis in the Belgian Congo', *Trans. R. Soc. trop. Med. Hyg.* **40**, 728–54.

HOOGSTRAAL, H. and DIETLEIN, D. R. (1964) 'Leishmaniasis in the Southern Sudan: recent results', *Bull. Wld Hlth Org.* **31**, 137–43.

HOOGSTRAAL, H. and HEYNEMAN, D. (1969) 'Leishmaniasis in the Sudan Republic', *Am. J. trop. med. & Hyg.* **18**, 1091–210.

KAMAL, A. M., ANWAR, M., MESSIH, G. D. and KOLTA, Z. (1947) 'Louse-borne relapsing fever in Egypt', *J. Egypt. Publ. Hlth Assoc.* **22**, 1–22.

KIRK, R. and LEWIS, D. J. (1947) 'Studies in leishmaniasis in the Anglo-Egyptian Sudan. IX. Further observations on the sandflies (*Phlebotomus*) of the Sudan; *Trans. R. Soc. trop. Med. Hyg.* **40**, 869–88.

KÖBERLE, F. (1968) 'Chagas disease and Chagas syndromes: the pathology of American trypanosomiasis' in *Advances in Parasitology*, vol. 6, London: Academic Press.

KOSTMANN, R., BARR, M., BENGTSSON, R., GARNHAM, P. C. C. and HALT, C. (1963) 'Kala-azar transferred by exchange blood transfusions in two Swedish infants', *Proc. 7th Internat. Cong. trop. Med. Malaria*, II, Rio de Janeiro.

LAINSON, R. (1969) and SHAW, J. J. 'Leishmaniasis in the Mato grosso', *Trans. Roy. Soc. trop. Med. Hyg.* **63**, 408–9.

LAINSON, R. SHAW, J. J. and LINS Z. C. (1969) 'Leishmaniasis in Brazil. III. The fox as a reservoir; *Trans. Roy. Soc. trop. Med. Hyg.* **63**, 741–5.

LAINSON, R. and STRANGWAYS DIXON, J. (1963) 'The epidemiology of dermal leishmaniasis in British Honduras', *Trans. R. Soc. trop. Med. Hyg.* **57**, 242–65.

LATYSHEV, N. I. and KRYUKOVA, A. P. (1941) 'On the epidemiology of cutaneous leishmaniasis', *Trudy voenno-med. Akad. RKKA.* **25**, 229–42.

LATYSHEV, N. I., KRYUKOVA, A. P., POVALISHINA, T. P. and CHERNYSHOV, V. I. (1947) 'Visceral leishmaniasis and jackal leishmaniasis in Southern Tajikistan', *Novost. Medizic.* **5**, 5–6.

LOW, G. C. and COOK, W. F. (1926) 'Congenital kala-azar', *Lancet* ii, 1209–1212.

LOW, G. C. and FAIRLEY, N. H. (1931) 'Observations on laboratory and hospital infections with yellow fever in England', *Brit. med. J.* i, 125–8.

LUMSDEN, W. H. R. (1952 and 1956) in *E. Afr. Virus Research Ann. Rep.*, Nairobi: Govt. Printers.

MANSON, P. (1899) *Tropical Diseases*, London: Cassell.

MANSON-BAHR, P. E. C. (1963) 'Active immunization in leishmaniasis', Ch. 17 in *Immunity to Protozoa*, Ed. P. C. C. Garnham, A. E. Pierce and I. Roitt, pp. 246–52, Oxford: Blackwell Scientific Publications.

MANZONI, A. (1969) *I Promessi Sposi*, London: Folio Soc.

MARINKELLE, C. J. (1966) 'Importancia de los murcielagos del tropico Americano en la salud publica', *Antioquio Med.*, **16**, 179–64.

MARTINI, G. A., SIMPSON, D. I. H. and ZLOTNIK, I. (1969) 'Vervet Monkey Disease', *Trans. R. Soc. trop. Med. Hyg.* **63**, 292–327.

MIGLIO, G.* (1959 and 1966) Larius, vol. 1 and 2. La città ed. il Lago di Como, Antologia.

MORRISON, R. J. G. (1937) 'Some cases of relapsing fever in Palestine', *J. R. Army med. Corps* **68**, 86–94.

MOSHKOVSKY, S. D. and DUHANINA, N. N. (1970) 'The geographical distribution and epidemiology of leishmaniasis, *Bull. Wld Hlth Org.* In press.

MULLIGAN, H. (1970) Ed. *The African Trypanosomiases*, with 27 contributors, London: George Allen & Unwin.

NADIM, A. and FAGHIH, M. (1968) 'The epidemiology of cutaneous leishmaniasis in the Isfahan province of Iran', *Trans. R. Soc. trop. Med. Hyg.* **61**, 534–42.

ORMEROD, W. E. (1963) 'A comparative study of the growth and morphology of strains of *Trypanosoma rhodesiense*', *Exp. Parasit.* **13**, 374–85.

PAVLOVSKY, E. N. (1966) *Natural Nidality of Transmissible Disease*, Ed. Norman D. Levine, Urbana: University of Illinois Press.

PENNA, H. A. (1934) 'Visceral leishmaniasis in Brazil', *Brasil-Med.* **46**, 950–53.

PETANA, W. B. (1969) 'Chagas's disease in British Honduras', *Trans. R. Soc. trop. Med. Hyg.* **63**, 9–10

PLAINFOSSE, B., HOEFFEL, J.-C., BRUMPT, L., LANGE, J.-C. and SERINGE, P. (1966) 'Un cas de trypanosomiase vraisemblablement congénitale chez une enfant de deux ans', *Méd. trop.* **74**, 0693–6

POLO, MARCO. (1968) *Travels*, Trans. R. Latham, London: Folio Soc.

REISS-GUTFREUND, R. J. (1967) 'The epidemiology of rickettsiosis on the Ethiopian plateau', *Am. J. trop. Med. Hyg.* **16**, 186–90.

RIOUX, J. A., GOLVAN, Y. J., CROSET, H. and HOUIN, R. (1969) 'Les leishmanioses dans le "midi" méditerranéen. Résultats d'une enquête écologique', *Bull. soc. Path. exot.* **62**, 332–7.

RIOUX, J. A., GOLVAN, Y. J., CROSET, H., HOUIN, R. and JUMINER, B. (1967) 'Epidemiology of the leishmaniases in the South of France', WHO, LEISHINF/67.10.

ROBERTSON, A. (1930) 'Note on a trypanosome morphologically similar to *Trypanosoma cruzi* Chagas, 1909 found in an opossum, *Didelphys marsupialis* captured at Tela, Honduras, Central America', *Ann. Rept. med. Dept. United Fruit Co.* 293.

RUBIO, M., GALECIO, R. and HOWARD, J. (1961) 'Dos casos de enfermedad di Chagas congenita', *Bol. Chil. Parasit.* **16**, 15–18.

SHEBA, C. (1968) 'Saul Adler. The scientist as consultant to the clinician: highlights of 30 years' association', *Israel J. med. Sci.* **4**, 1053–6.

SICI, A. and BROCKEN, L. (1940) 'Les manifestations sporadiques du typhus amaril au Soudan français, et leur expression épidémiologique', *Bull. Soc. Path. exot.* **33**, 266–71.

SOUTHGATE, (1967) '*Leishmania adleri* and natural immunity', *J. trop. Med. Hyg.* **70**, 33–36.

STRODE, G. K. (1951) *Yellow Fever*, New York: Hill Book Co.

SWELLENGREBEL, N. H. (1913) *Rep. Ned. Ind. med. Civil Service* 2, Part 1.

SYMMERS, W. ST C. (1960) 'Leishmaniasis acquired by contagion. A case of marital infection in Britain', *Lancet* **i**, 127–30.

TAYLOR, R. M., WORK, T. H., HURLBUT, H. S. and RIZK, F. (1956) 'A study of the ecology of West Nile virus in Egypt', *Am. J. trop. Med. Hyg.* **5**, 579–620.

WIJERS, D. J. B. (1969) 'The possible role of domestic animals in the epidemiology of *T. rhodesiense* sleeping sickness', *Bull. Soc. Path. exot.* **62**, 334.

WORK, T. II. (1958) 'Kyasanur Forest Disease', *Proc. 6th Internat. Congr. trop. Med. Malaria* Lisbon **5**, 180–96.

* The author acknowledges with gratitude his debt to John Marshall, Rockefeller Foundation, for drawing his attention to the original sources.

4

MALARIA AS A MEDICAL AND VETERINARY ZOONOSIS

THE CONCEPT of a zoonosis should include not only infections of man acquired from vertebrate animals, but also infections of *domestic* animals acquired from wild animals (see p. 5 above). Both medical and veterinary zoonoses are governed by the same general principles and show similar dynamic changes from the primaeval state in natural foci to eventual extinction with urbanization. The zoonotic course of veterinary trypanosomiasis has already been referred to (p. 27). In this chapter, malaria is considered in this particular context; the human forms represent an obsolescent zoonosis, while the veterinary forms are mostly on the crest of the wave.

The present status of human malaria is thus almost identical with Indian kala-azar in which the animal reservoir has entirely disappeared and the insect vector is largely domestic in its habits; the status of veterinary malaria resembles that of Iranian cutaneous leishmaniasis, where the zoonosis is flourishing in man and gerbil living side by side.

HUMAN ZOONOSIS

The evolution of human malaria may be considered first, and although the following reconstruction cannot be proved, it fits in with the few known facts. The possible evolution of the parasites and their hosts has been discussed in detail by the author (1963).

The vertebrate hosts comprised lower monkeys, the apes, the prehominids and early man of the Old World, and the lower monkeys (and man) of the New World.

It is thought that malaria parasites existed only in the Old World animals, presumably first in the oriental monkeys, which still show an extensive range of parasites. Sergiev and Tiburskaya (1967) assume that a zoonosis involving *Plasmodium cynomolgi bastianellii*, macaques and early man in South-east Asia, evolved into the human *Plasmodium vivax*, which with migrations to the north from the tropics changed its characters to form the various subspecies which are known today (e.g. *hibernans*, St Elizabeth, etc.) The apes (gibbons and orang utans) of this region also harboured a variety of malaria parasites.

In Africa, malaria was absent from the vast monkey population except for the rare *Plasmodium gonderi* of mangabeys and mandrills of the West African coast; the apes (chimpanzees and gorillas) however were heavily infected with three species closely resembling the human parasites of today.

We may now speculate on the nature of the pristine zoonosis in the Old World. The arboreal prehominids and early man, monkeys and apes lived together in the forests of tropical Asia and Africa. Sylvatic species of *Anopheles* bit these animals indiscriminately and malaria parasites passed to and fro between them. In the Lower Pleistocene Age, some men began to leave the forest for the savannah; they were accompanied by their malaria parasites of simian origin, and these organisms had to find new vectors, viz. non-sylvatic anophelines, in order to continue their existence. As man began to live in caves, or other primitive shelters, both in the forest and savannah, the mosquitoes followed him indoors, and domestic strains or species of these insects evolved. The next stage in human evolution was accompanied by further migrations, which brought man and his parasites into contact with other species of *Anopheles*. In this way, segregation of the parasites occurred, to be followed in due course by speciation and the appearance of special types of *Plasmodium* peculiar to man.

At an early stage of evolution, the zoonosis became extinct, except perhaps in those forest dwellers who continued to live next door to the simian population and at first shared their parasites.

Man eventually became cosmopolitan and far removed from the monkeys and apes; the latter on the other hand remained segregated in their original environment with its specialized conditions and parasites.

In contrast to the Old World, the situation in the Americas is very different; no apes exist and man is of recent introduction—monkeys however appeared even earlier than in the Old World and speciation of these animals is more prolific. Malaria parasites are confined to two organisms—the widespread *Plasmodium brasilianum* which causes quartan malaria in many species of monkeys in South and Central America and the rare *P. simium* of southern Brazil.

An intriguing suggestion is that infections of *P. brasilianum* represent a zoonosis in reverse; the parasite is thought to have been introduced by the human immigrants from the Old World, either in historical times or more probably at a remote epoch. These people were accompanied by the notoriously long-lived *P. malariae*, which in due course became adapted to one species of monkey and then to many others. The evidence for this theory is firstly that *P. brasilianum* is readily transmissible to man by mosquito bite (Contacos and Coatney, 1963) with the production after several human passages of typical quartan malaria; it would be interesting to observe the behaviour of human strains of *P. malariae* in South American monkeys, for we know that the owl monkey and marmosets are easily infected with other species of human *Plasmodium*. Secondly *P. brasilianum* bears a remarkable morphological similarity to *P. malariae* in its primitive exoerythrocytic stages. The schizont is exceptionally large and causes gross hypertrophy of the host cell nucleus, its periphery is occupied by orange- and pink-staining, 'bubbly' vacuoles and it takes more than 12 days, probably 15, to reach maturity. All these features are shared by the human *P. malariae* (Lupascu *et al.*, 1967). On the contrary, the quartan malaria parasites of the Old World monkeys have quite a different tissue stage: the schizonts are the smallest (instead of the largest) of any of the mammalian species, undergo characteristic cytomeric differentiation and take only 11 or 12 days to develop (Garnham, 1951).

Although the zoonotic origin of human malaria is speculative, the relative susceptibility of man to simian parasites, and of apes and monkeys to human parasites have been repeatedly demonstrated. Thus man can be infected by mosquito bite with the following species of simian and chimpanzee parasites:

P. cynomolgi bastianelli: Eyles *et al.* (1960)
P. cynomolgi cynomolgi, M. Strain: Schmidt *et al.* (1961)
P. cynomolgi, Cambodian strain: Bennett and Warren (1965);
  Coatney *et al.* (1961)
P. shortti: Coatney *et al.* (1966)
P. brasilianum: Contacos *et al.* (1963)
P. knowlesi. Chin *et al.* (1965)
P. schwetzi: Coatney (1968)

Man can be infected by blood containing the following species of simian and higher ape parasites:

P. inui: Dasgupta (1938)
P. eylesi: Warren *et al.* (1965)
P. knowlesi: Ciuca *et al.* (1955)
*P. malariae: Rodhain (1940)
P. schwetzi: Rodhain and Dellaert (1955)

The susceptibility of apes and monkeys to human parasites is best revealed by splenectomy, though a low parasitaemia may sometimes be established in intact animals, while exoerythrocytic schizogony is demonstrated in intact chimpanzees without difficulty. After frequent passage of the human parasites in animals, some degree of adaptation occurs and splenectomy becomes unnecessary. The following combinations have been successfully applied:

P. vivax produces light infections in chimpanzees which become heavy after splenectomy. Many unsuccessful attempts have been made to infect splenectomized rhesus monkeys with this parasite; yet South American monkeys have proved to be susceptible: *Aotus trivirgatus*, with and without spleens, were infected by blood and sporozoites (Porter and Young, 1966), *Saguinus geoffroyi* (Porter and Young, 1966) and splenectomized *Saimiri sciurus* (Deane *et al.*, 1966).

P. ovale readily infects chimpanzees which have been splenectomized, but intact animals are only feebly susceptible (Bray, 1957).

P. malariae is a natural parasite of the chimpanzee which can less easily be infected with human strains of the parasite. This

* chimpanzee strain

suggests that a minor degree of speciation of the two forms of
quartan malaria has occurred; moreover, Bray (1960) showed
that the vector (*A. gambiae*) of the human strain was insusceptible to the one in the chimpanzee.

P. falciparum was found by Taliaferro and Taliaferro (1934) to
be able to infect baby *Alouatta palliata* if given in a heavy dosage,
and later by Cadigan *et al.* (1966) to infect various species of
macaques (including rhesus) which had been splenectomized.
The New World owl monkey (*Aotus trivirgatus*) is also susceptible.
Splenectomized chimpanzees are easily infected both with the
blood and sporozoite forms of the parasite (Bray and Gunders,
1962) as are gibbons (Gould *et al.*, 1966).

In some of the above examples, even though parasitaemia
may be heavy, gametocytes are either absent or are non-infec-
tive to mosquitoes, indicating that the exotic host is not entirely
suitable. This is one of several reasons why such experimental
procedures do not necessarily confirm the idea that human
malaria is a zoonosis in nature.

The first indication that monkey malaria is a potential zoonosis was the occurrence of several cases of laboratory infections
of *P. cynomolgi bastianellii* in the United States (Eyles *et al.*,
1960). The workers had been handling large batches of mosquitoes infected with this parasite, and were accidentally bitten.
Several other episodes occurred in the United States, and also
in the writer's department in London. The infectivity of this

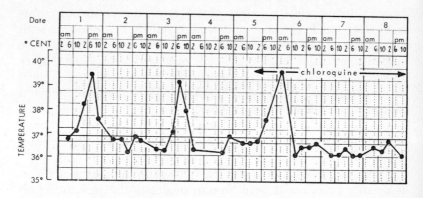

FIG. 10. Temperature chart of Mrs K. Monkey malaria (*P. cynomolgi*).

parasite and of other subspecies in such circumstances is marked, and diagnosis of the resultant illness may sometimes be difficult, as in the most recent case at the London School. A large batch of *Anopheles stephensi* infected with *P. cynomolgi cynomolgi* was dissected in a room in the department, protected with mosquito netting, which inevitably had to be opened a few times in the course of the work. On one such occasion, an infected mosquito must have escaped and flown some distance down the corridor to enter another room occupied by a secretary. Fifteen days later this lady (Mrs K) became ill with headache and fever, which was thought to be influenza. She got worse and was admitted to hospital. As she had been working at the 'Tropical School', malaria was half-jokingly suggested as the diagnosis, but blood films proved at first to be negative. Nevertheless, the temperature chart (Fig. 10) showed a typical tertian periodicity and the patient's symptoms (including splenomegaly) worsened. At last, on the fifth day after admission, very scanty vivax-like parasites were found in the thick film, and when a sample of her blood was inoculated into a rhesus monkey, the animal developed a typical infection of *P.c. cynomolgi* malaria. The patient rapidly recovered after the administration of chloroquin. This case illustrates all the typical features of simian malaria in man, which are as follows:

1. Misdiagnosis
2. Long incubation period
3. Low density of parasitaemia
4. Symptoms unexpectedly severe
5. Rapid cure with anti-malarial drugs.

The occurrence of these laboratory infections suggested that malaria in certain regions must be a zoonosis in nature, and a large-scale investigation was started immediately by Don Eyles, Coatney and their colleagues in Malaya, the home of monkey malaria. The blood from indigenous people living in the jungle and suffering from malaria was inoculated into rhesus monkeys, but always with negative results. Thus Warren *et al.* (1969) tested 2000 such people in an area where *P. knowlesi* was common in the monkeys and mosquitoes (*A. leucosphyrus*), and 300 people in another place where the vector (*A. balabacensis*) readily invades the villages; no simian malaria parasites were found in these people.

In spite of the negative results of this intensive investigation and of the failure of malaria to recur in the few places where it had been eradicated in the human but not the simian population (e.g. Taiwan), the existence of a natural reservoir of the disease remains a distinct possibility. And this has now been confirmed by the occurrence of two actual cases, one emanating from Malaya and the other from South America.

The first instance was reported by Chin *et al.* (1965) and concerned a surveyor, returned from the Malayan jungle to the United States, ill with an unknown fever; in Maryland rings of what was thought to be *P. falciparum* were found; this locality was the headquarters of the Malayan monkey malaria team and it is not surprising that directly the investigators heard about the case, they inoculated the man's blood into a rhesus monkey. In 6 days the monkey was dead with an overwhelming infection of *P. knowlesi*.

The second case occurred in an entomologist who had been collecting mosquitoes on a tree platform in a small forest on the outskirts of São Paulo (Deane *et al.*, 1966). This is one of the few localities where *P. simium* is known to occur in the howler monkeys. The man became ill with a tertian fever and vivax-like parasites were found in small numbers in his blood; samples of the latter were inoculated into a splenectomized *Saimiri* monkey which developed a typical infection of *P. simium*.

With these facts before us, it is possible to assess their significance for malaria as a zoonosis, and the writer (Garnham, 1967) has recently recapitulated the evidence for and against this theory, which briefly may be summarized as follows:

Evidence for malaria as an important zoonosis

1. Two proven cases have been found in nature (Pahang and São Paulo).

2. Accidental laboratory infections in man of simian malaria of a mosquito origin are not uncommon.

3. Six species of simian malaria parasites have been transmitted to man by mosquito bite and cyclical passage has been maintained with some increase in virulence.

4. Some sylvatic anopheline vectors of malaria bite man and monkey indiscriminately.

5. The known transformation of jungle yellow fever into urban yellow fever offers a parallel to what might happen in malaria.

Evidence against malaria as an important zoonosis

1. Cases are rare and deliberate searches in Malaya, Ceylon and Brazil have yielded negative results.

2. Many vectors of simian malaria do not feed on man, and common man-biters often cannot transmit monkey or ape parasites.

3. Simian malaria parasites give rise at first to low infections in man with scanty or absent gametocytes, so they are unlikely to infect mosquitoes.

4. Malaria (of a simian origin) does not appear to have broken out in places where the human disease has been eradicated.

Most malariologists today minimize the significance of the zoonosis problem in malaria. Bruce-Chwatt (1966) concludes that its practical importance is very slight; Bray (1968) feels that more investigations are necessary, but thinks that the possibility of the transmission of quartan malaria from chimpanzees to man is remote; Deane (1968) states that the existence of monkey malaria constitutes no real danger to eradication programmes, and the study group of WHO (1968) on the parasitology of malaria reached identical conclusions.

The final event in the zoonotic chain of malaria, as in other infections, is the disappearance of the domestic vector and the direct transference of the parasite from mother to foetus via the placenta. This mechanism of transmission, though uncommon, occurs in all four species of human *Plasmodium*.

VETERINARY MALARIA

We have seen that a feral reservoir plays little part today in the epidemiology of human malaria. In avian malaria on the con-

trary, the disease in domestic birds is often a full scale zoonosis, i.e. closely linked with infections in wild birds. Three
forms of bird-malaria are of some veterinary importance, viz.
P. gallinaceum and *P. juxtanucleare* in chickens, and *P. durae* in
turkeys, while malaria *sensu latu* affects chickens in the form of
Bangkok haemorrhagic fever, due to *Leucocytozoon* (= *Akiba*)
caulleryi and ducks and geese with the equally virulent *Leucocytozoon simondi*. All these zoonoses are examples of malaria in
domestic birds, acquired from local wild birds; it is clear that
the disease could not have been peculiar to domesticated birds,
because it breaks out in places where the bird is not indigenous.
[§]

Plasmodium gallinaceum

The natural host of *P. gallinaceum* is the jungle-fowl of South-east
Asia, which occurs in the form of different species or subspecies
of *Gallus* in this area. The parasite was found in 1 out of 40
specimens of *G. sonneratii* in Madras by Shortt *et al.* (1941); *G.
lafayettei* is the probable reservoir in Ceylon. The infection is of
low density in these birds, though little is known about the
experimental course of infection in them. The jungle fowl is the
ancestor of the domestic fowl, and the malaria parasites of these
two birds (wild and domestic) offer an interesting parallel to
those found in the higher primates and man: the avian species
have scarcely had time to diverge from each other, because
domestication of their hosts is of recent origin (*c.* 5000 years);
the primate species separated probably 50 000 years ago and
speciation has become possible. The local chickens (*G. gallus*)
contract the infection as a veterinary zoonosis, but the disease is
not severe and most birds recover. Exotic breeds, however, have
little natural immunity and severe epizootics may decimate
them. The birds die from cerebral malaria, due to blockage of
the cerebral capillaries by exoerythrocytic schizonts in the
endothelium.

Epizootics of *P. gallinaceum* are practically only seen in countries where the jungle-fowl occurs, i.e. in India, Ceylon, Indo-
China, Sumatra, the Philippines and Malaya. Outbreaks have
been reported from Egypt and West Africa, but the only valid
epizootic outside Asia was that reported by Krishnamurti *et al.*

(1961) in the United States, and this infection probably arose from a laboratory strain of the parasite (or was falsely reported).

The natural vector of *P. gallinaceum* in Ceylon was discovered by Niles *et al.* (1965) to be *Mansonia crassipes*; but many mosquitoes act as efficient vectors in the laboratory and the progress of an epizootic in semi-urban areas suggests that a more domestic vector like *Aedes aegypti* may be concerned. In other words, the veterinary zoonoses has expired and the wild vector has been replaced by a domestic one exactly like a medical zoonosis.

Some inhibitory factor must exist which interferes with the wide dispersal of *P. gallinaceum*, and it seems that the presence of the feral host is necessary for the continued survival of epizootics.

Plasmodium juxtanucleare

This tiny and easily overlooked parasite of chickens has a much more extensive distribution than *P. gallinaceum*, though the natural host is apparently again the jungle-fowl (*G. lafayettei*); 8 of these birds were examined by Dissanaike (1963) in Ceylon and one was found to harbour a light infection.

Epizootics of *P. juxtanucleare* malaria are common in Ceylon and Malaya and probably represent direct zoonoses, transmitted by *Culex sitiens* (Bennett *et al.*, 1963) in a unique sporogonic cycle involving pedunculated oocysts.

P. juxtanucleare infections are widespread in the tropics and subtropics, for this species is an exception and does not always need the original host for its maintenance, and easily dispenses with the zoonotic link. It is thus found in countries where the genus *Gallus* is absent in the wild state, as in the Americas. The parasite was actually first found in the domestic fowl in Brazil; then in Mexico and later in Uruguay. It seems probable that this long-lived species of parasite has accompanied its secondary host—the domestic hen—from the original habitat in South-east Asia, to Taiwan and Japan, and perhaps in pre-Colombian days to the New World. It is strange that the companion species, *P. gallinaceum*, with a much wider range of invertebrate hosts, and of the same geographical origin has been unable to undergo similar migrations.

Plasmodium durae

This parasite has a distribution restricted apparently to a small area in the Kenya Highlands, although a few doubtful reports come from elsewhere in Africa. *P. durae* gives rise to severe epizootics in turkeys, and it was at once realized that the infections must represent a zoonosis, as turkeys are not indigenous to the Old World. The feral counterpart of this turkey parasite was sought in vain for many years; its morphology is sufficiently distinctive for other malaria parasites of game birds (e.g. *P. fallax* of the guinea-fowl) to be excluded.

Finally in 1967, Southgate solved the mystery by isolating a parasite from the yellow-necked francolin (*Pternistes leucoscepus*) which presents an appearance identical with *P. durae* in the blood and tissues and gives rise to typical infections in turkeys. The restricted distribution of this francolin, with its special parasite probably limits turkey malaria to those few places in Africa where the two birds overlap. Only then can the zoonosis arise; presumably, too, other factors in the biocenosis must be present such as a species of mosquito capable of transmitting the parasite and biting both types of birds. *Culex pipiens* meets these conditions; transmission is easy and frequent epizootics occurs in the turkeys in the semi-domestic surroundings in which these birds are maintained. The natural mosquito host (*C. univittatus*) and the feral reservoir both disappear under these conditions.

Leucocytozoon

L. (= *Akiba*) *caulleryi* causes a highly fatal disease in chickens in Thailand, Ceylon, Malaya, Japan and Korea, but hitherto its feral counterpart has not been found although the Ceylon jungle-fowl was suspected. It is impossible, therefore, to discuss the zoonotic aspects of this important veterinary infection.

The natural history of *L. simondi* however is well-known and provides an interesting example of a zoonosis directly due to human agency. This haemosporidian parasite is common in wild ducks and geese in many parts of the world, including Asia, Europe and North America, and it is transmitted by the bite of various species of *Simulium*. If the infection is introduced into a flock of ducklings, the birds may be normal in the morning, ill in the afternoon and dead the following day (Levine, 1961). The disease in these birds constitutes an alarming veterinary zoonosis in the northern parts of the United States and Canada.

In order to improve the nutritional status of the inhabitants of a subarctic region in Canada, geese were introduced in 1953 (Levine and Hansen); the project was a failure, because of migrating wild duck which arrived in the early summer with gametocytes of *L. simondi* in their blood; *Simulium* spp. were present in enormous numbers, these insects fed on the wild duck and transmitted the parasite to the domestic geese, which suffered an excessively high mortality from the infection.

In human malaria, almost all infections have long passed the zoonotic stage and the actual occurrence of a zoonosis in nature has only been demonstrated twice; in avian malaria, most infections are still active zoonoses though it may be difficult to discern the exact pathway of the infection. It is probable that in both examples, human and veterinary, the extermination of the vector or protection from its bite offers the best prospects of control.

REFERENCES

AKIBA, K. (1964) 'Leucocytozoonosis in Japan', *Bull. Off. int. Epizoot.* **62**, 1017–22.

BENNETT, G. F., EYLES, D. E., WARREN, M. and CHEONG, W. H. (1963) '*Plasmodium juxtanucleare* a newly discovered parasite of domestic fowl in Malaya', *Singapore med. J.* **4**, 172–3.

BENNETT, G. F., WARREN, M. and CHEONG, W. H. (1965) 'Strains of *Plasmodium cynomolgi* identified by characteristics of the sporogonic cycles', *Singapore med. J.* **5**, 15–22.

BRAY, R. S. (1960) 'Studies on malaria in chimpanzees. VIII. The experimental transmission and pre-erythrocytic phase of *Plasmodium malariae* with a note on the host range of the parasite', *Am. J. trop. Med. Hyg.* **9**, 455–65.

BRAY, R. S. (1968) 'Zoonotic Potential of Blood Parasites' in *Infectious blood diseases of man and animals,* I, New York: Academic Press.

BRAY, R. S. and GUNDERS, A. E. (1962) 'Studies on malaria in chimpanzees. IX. The distribution of the pre-erythrocytic forms of *Laverania falcipara*', *Am. J. trop. Med. Hyg.* **11**, 437–9.

BRUCE-CHWATT, L. J. (1966) 'Malaria zoonosis in relation to malaria eradication', *Trop. geogr. Med.* **20**, 50–87.

CADIGAN, F. C., SPERIZEL, R. O., CHAICUMPAN, V. and PUKOM-
CHAREON, S. (1966) '*Plasmodium falciparum* in non-human primates',
Milit. Med. **131**, 959–61 (Suppl.).
CASSAMAGNAGLI, A. (1947) 'Malaria en las aves del Uruguay' *Facultad de
Veterinaria, Montevideo*, p. 93.
COATNEY, G. R. (1968) 'Simian malaria in man: facts, implications and
predictions', *Am. J. trop. Med. Hyg.* **17**, 147–55.
COATNEY, G. R., CHIN, W., CONTACOS, P. G. and KING, H. K. (1966)
'*Plasmodium inui*, a quartan-type malaria parasite of Old World monkeys
transmissible to man', *J. Parasit.* **52**, 660–3.
COATNEY, J. R., ELDER, H. A., CONTACOS, P. G., GREENLAND, R.,
ROSEAU, R. N. and SCHMIDT, L. H. (1961) 'Transmission of the M
strain of *Plasmodium cynomolgi* to man', *Am. J. trop. Med. Hyg.* **10**, 673–8.
CONTACOS, P. G. and COATNEY, G. R. (1963) 'Experimental adaptation
of simian malarias to abnormal hosts', *J. Parasit.* **49**, 912–18.
DASGUPTA, B. M. (1938) 'Transmission of *Plasmodium inui* to man', *Proc.
natn. Inst. Sci., India* **4**, 241–4.
DEANE, L. M., FERREIRA-NETO, J. and SILVEIRA, I. P. S. (1966)
'Experimental infection of a splenectomized squirrel monkey with
Plasmodium vivax', *Trans. R. Soc. trop. Med. Hyg.* **60**, 811–12.
DHANAPALA, S. B. (1962) 'The occurrence of *Plasmodium juxtanucleare*
Versiani and Gomes, 1941 in domestic fowls in Ceylon', *Riv. Malariol.* **41**,
39–46.
DISSANAIKE, A. S. (1963) 'On some blood parasites of wild animals in
Ceylon', *Ceylon vet. J.* **II**, 75–86.
EYLES, D. E., COATNEY, G. M. and GETZ, M. E. (1960) 'Vivax-type
malaria parasite of macaques transmitted to man', *Science, N.Y.* **132**,
1812–13.
GARNHAM, P. C. C. (1951) 'The mosquito transmission of *Plasmodium inui*
Halberstädter and Prowazek, and its pre-erythrocytic development in the
liver of the rhesus monkey', *Trans. R. Soc. trop. Med. Hyg.* **45**, 45–52.
GARNHAM, P. C. C. (1963) 'Distribution of simian malaria parasites in
various hosts', *J. Parasit.* **49**, 905–11.
GARNHAM, P. C. C. (1967) 'Malaria in Mammals excluding Man' in
Advances in Parasitology, vol. 5, London: Academic Press.
GARNHAM, P. C. C. (1969) 'Malaria parasites as medical and veterinary
zoonoses', *Bull. Soc. Path. exot.* **62**, 325–32.
GARNHAM, P. C. C., BAKER, J. B. and NESBITT, P. E. (1963) 'Transmis-
sion of *Plasmodium brasilianum* by sporozoites, and the discovery of an
exoerythrocytic schizont in the monkey liver', *Parasitol.* **5**, 5–9.
GOULD, D. J., CADIGAN, F. C. and WARD, R. A. (1966) 'Falciparum
malaria: Transmission to the gibbon by *Anopheles balabacensis*', *Science* **153**,
1384.
KRISHNAMURTI, P. V., PEARDON, D. L., TODD, A. C. and McGIBBON,
W. H. (1961) 'Natural occurrence of a *Plasmodium* in chickens of domestic
fowl in Malaya', *Singapore med. J.* **4**, 172–3.

LEVINE, N. D. (1961) 'Protozoan parasites of domestic animals and of man', Minneapolis: Burgess Publicity Co.

LEVINE, N. D. and HANSEN, H. C. (1953) 'Blood parasites of the Canadian goose', *J. Wildl. Mgmt.* **17**, 185–95.

LUPASCU, GH., CONSTANTINESCU, P., NEGULICI, E., GARNHAM, P. C. C., BRAY, R. S., KILLICK-KENDRICK, R., SHUTE, P. G. and MARYON, M. (1967) 'The late primary exoerythrocytic stages of *Plasmodium malariae*', *Trans. R. Soc. trop. Med. Hyg.* **61**, 482–9.

NILES, W. J., FERNANDO, M. A. and DISSANAIKE, A S. (1965) '*Mansonia crassipes* as the natural vector of filarioids, *Plasmodium gallinaceum* and other plasmodium of fowls in Ceylon' (Correspondence), *Nature, Lond.* **205**, 411–12.

PORTER, J. A. and YOUNG, M. D. (1966) 'Susceptibility of Panamanian primates to *Plasmodium vivax*', *Milit. Med.* **131**, 952–7.

RODHAIN, J. (1950) 'Les plasmodiums des anthropoïdes de l'Afrique centrale et leur rélations avec les plasmodium humains. Réceptivité de l'homme au *Plasmodium malariae* (*Plasmodium rodhaini* Brumpt) du chimpanzé', *C. r. Séanc. Soc. Biol.* **133**, 276–7.

RODHAIN, J. and DELLAERT, R. (1955) 'Contribution à l'étude de *Plasmodium schwetzi* E. Brumpt (3 note). L'infection à *Plasmodium schwetzi* chez l'homme', *Ann. Soc. belge Méd. trop.* **35**, 757–76.

SCHMIDT, L. H., GREENLAND, R. and GENTHER, C. S. (1961) 'The transmission of *Plasmodium cynomolgi* to man', *Am. J. trop. Med. Hyg.* **10**, 679–88.

SERGIEV, P. G. and TIBURSKAYA, N. A. (1967) 'A propos du l'évolution du *Plasmodium vivax*', *Arch. roum. Path. exp. Microbiol.* **26**, 473–84.

SHORTT, H. E., MENON, K. P. and IYER, P. B. S. (1941) 'The natural host of *Plasmodium gallinaceum* (Brumpt, 1935)', *J. Malar. Inst. India* **4**, 175–8.

SOUTHGATE, B. A. (1967) in *Annual Report, Division of Insect-Borne Diseases, Kenya*, J. M. D. Roberts, pp. 27–28.

TALIAFERRO, W. H. and TALIAFERRO, L. G. (1934) 'The transmission of *Plasmodium falciparum* to the howler monkey, Alouatta sp. I and II', *Am. J. Hyg.* **19**, 318–42.

WARREN, M., BENNETT, G. F., SANDOSHAM, A. A. and COATNEY, G. R. (1965) '*Plasmodium eylesi* n. sp. a tertian malaria parasite from the white-handed gibbon, *Hylobates lar*', *Ann. trop. Med. Parasit.* **59**, 500–8.

5

THE CHANGING PATTERN OF PARASITIC DISEASES AS A PUBLIC HEALTH PROBLEM

THE EARLIER chapters of this book are devoted to the general principles which govern the pattern of parasitic diseases, and to some specific examples of how their course varies under natural conditions or at least under circumstances which have not been deliberately altered by human activities. This chapter is concerned with the changes which take place as the result of urbanization and development of the country, including the effects of various public health campaigns. In discussions on this subject, it is impossible to ignore the general effect of advancing civilization, a picture which is Janus-faced, one pleasing and the other horrifying; although one face may be that of Dr Jekyll, the other is that of Mr Hyde. The most obvious example of the latter is the spectre of over-population, which haunts the politician with its vision of anarchy, and the humanitarian with the prospect of malnutrition and general misery.

In temperate countries, the recent changes in parasitic diseases have almost always been for the better; malaria has gone, the fly-borne intestinal infections are greatly diminished, cholera, yellow fever, plague, relapsing fever and typhus are diseases of the past. Most arthropod-borne virus diseases are under control and, although helminthic infections are still prevalent, they are not in general a serious problem.

In the tropics and subtropics, the picture is very different. Chagas' disease undermines the health of the population over much of Brazil and other parts of South America; the ravages of schistosomiasis in most parts of the tropics are becoming increasingly apparent, even malaria in some places is as prevalent as ever, while the intestinal helminths continue to exert a damaging effect on the general health of hundreds of millions.

Yet, some tropical diseases, like yaws, have practically been abolished by well organized campaigns of penicillin administration, and it would be quite true to claim that the *initiated* traveller could, by appropriate measures and with a little luck, avoid infection with all the parasitic diseases of warm climates.

IMPORTANT FACTORS IN PUBLIC HEALTH
IN THE TROPICS

There are five factors which have a profound overall influence on public health in tropical countries. These are (1) food supply, (2) land development, (3) political organization, (4) communications and (5) education. Each of these factors, except perhaps the last, operates in two ways, one beneficial and the other harmful. A few examples follow.

Food supply

Good nutrition confers some protection against most parasitic diseases, e.g. *Entamoeba histolytica* is much less invasive in people living on a mixed diet, which includes meat, than in those individuals restricted to carbohydrates (Elsdon-Dew, 1949) but, strangely enough, malaria is relatively mild during famines— starvation of the host is accompanied by malnutrition of the parasite which is unable to multiply at the normal rate.

Land development

Land development drives away animal reservoirs of disease, such as gerbils or antelopes, and it gets rid of dangerous vectors like the tsetse flies. But many bush-clearing operations in the past have ultimately failed because there has not been a sufficient population to occupy the land that the tsetses have vacated. Agricultural measures are directed to improvements in food supply, but at the same time, the construction of irrigation canals, swamp drainage, dams, rice fields, etc., may easily be accompanied by an upsurge in the population of bilharzial snails and of malarial mosquitoes.

Political organization

The political control of most of tropical Africa by the great

European powers in the first half of this century produced an organized society based on good education, better housing, fine health services, and general stability. Barbarism and internecine strife were replaced by civilization, though civilization in its turn introduced *new* diseases, like measles and whooping-cough, and intensified in its cities *old* ones, like tuberculosis. But with the advent of independence, there has been in some parts a decline in standards and civil war. A grim resurgence in the major infectious diseases like trypanosomiasis, has followed 'decolonization'. Any political disturbances are apt to lead to a heightened incidence of parasitic disease; thus Forrester and Nelson (1965) describe how in the Mau Mau emergency the youth went wild on Mount Kenya, ignored the ancient tribal taboo regarding the consumption of the flesh of certain animals and devoured uncooked wild pig: a severe outbreak of trichinellosis, with many deaths, was the result.

Communications

The fourth factor which has influenced the incidence of parasitic diseases is the introduction on a wide scale of roads, railways and air routes. Communicable diseases are spread by increased communications, and many examples are mentioned in these lectures, e.g. malaria, sleeping sickness, yellow fever, plague and cholera. Even in early days, infections were disseminated by travel. Thus, tick-borne relapsing fever used to be prevalent in camps along the safari routes in Tanganyika and into Kenya and Uganda. Porters and recruited labourers travelled hundreds of miles and were accompanied by *Ornithodorus moubata* in their packs and *Borrelia duttoni* in their blood. The labour camps were a permanent source of infection from which the spirochaete spread into the neighbouring districts. Another parasitic infestation, the chigger flea, was brought into Africa by travellers from South America, perhaps as early as the eighteenth century, but for certain in 1872, when an English ship brought a person with chiggers in his feet from South America to Angola (Hoeppli, 1963). The infection spread rapidly in West Africa, but did not reach the East Coast until twenty years later; thence the parasite travelled to Zanzibar and Madagascar and eventually entered Bombay.

The danger of a disease like yellow fever invading Asia was well-recognized and international action to prevent such a disaster was taken by requiring travellers to be immunized against the disease, aircraft to be sprayed with an insecticide on arrival and aerodromes to be rendered *Aedes* free. Possibly because of these precautions, or more probably due to some unknown factor, yellow fever has never been established in Asia.

In general the international sanitary regulations are no longer adequate (Dorolle, 1968). Jet-age epidemiology has made nonsense of the outdated idea of quarantine, which according to the Deputy Director General of WHO, implies the existence of a 'sin that spoils the national image and must be hidden away as soon as possible.' Dorolle concludes that a scientific approach should replace the legalistic measures, but realizes the difficulties, e.g. how is it possible to prevent the introduction of a person with malaria into a place where the disease has been eradicated but where the vector still exists?

Maegraith (1968) stressed how the global pattern of disease is being re-arranged by the enormous number of travellers and immigrants who move from one continent to another or from province to province within a country at a speed much faster than the incubation period of infectious diseases. Moreover, the parochial physician often fails to diagnose the exotic disease from which the newly arrived traveller is suffering. Thus, it is important to ask the patient, 'where have you been?' if there is any doubt about his origin.

Education

The benefits of improved communications are too obvious to mention as they affect all major human activities, and the same applies to the last factor, education. It has already been said (p. 63) that the 'initiated' could avoid contracting practically any tropical infection. But initiation entails education, and this is far from universal in the tropics. One feels so powerless when confronted by the lethargy or ignorance of the inhabitants of places where schistosomiasis is rife, and by their failure to introduce the simple sanitary measures which could immediately break the cycle of transmission of the trematode worm. Instead, the local authorities attempt to apply—often with the greatest ingenuity—the much less effective procedures for the destruction of the snail vector or of the parasite in the human host. In an educated community, the disease would be made to disappear very quickly by the adoption of ordinary hygienic routines. Webbe (1969) who has had much experimental success by the use of molluscicides, points out that 'socio-economic levels must be raised considerably in many endemic areas before improvements in sanitation will be accepted and considered desirable, and therefore achieve the control of schistosomiasis that the efficient disposal of excreta theoretically should'.

It is regrettable that the lessons given in the late Dr A. R. Paterson's *Book of Civilization* have not been more rammed home; the book however has been translated into several African languages.

The following examples are taken largely from Africa, in order to illustrate the changes in the pattern of parasitic diseases which are emerging in that continent today (Garnham, 1968). Their effect on health is difficult to assess, particularly in rural populations where multiparasitism is practically the fate of everyone. Walker (1963) truly states that only long-term studies, accompanied by adequate controls in which the variables are altered singly, will demonstrate the real effect of a parasite. Malaria is often said to have been responsible for the decline in some of the ancient civilizations and for the difficulties which confront the emerging ones of today, and in an attempt to confirm this theory, Dr and Mrs Wilson (1960 and 1962) watched the changes in a large population in northern Tanzania where malaria was being eradicated, and compared them with a control zone which was left untouched. The physiologists were consulted on how to measure health, but no precise answer could be extracted from them, and observations were confined to (a) physical performance, (b) growth (height and weight), (c) anaemia and (d) simple psychological tests, which had to be abandoned owing to the non co-operation of the inhabitants. The results were almost impossible to interpret, though Dr and Mrs Draper (1960) showed that the infantile mortality rate was halved and there was a rise of 2 gr per cent in the haemoglobin level in the infants; but no specific effect attributable to malaria control was demonstrated in the pattern of growth of the children.

Some surveys are content to rely on measurements of morbidity rather than of health; these indices in principle are simpler, e.g. absenteeism in labour forces, school attendances, etc., but they represent indirect evidence which is notoriously unsatisfactory.

The major difficulty in trying to assess the results of malaria eradication campaigns are two: (1) the removal of an infection like malaria sets up a chain reaction, whereby anaemia is lessened because there are no parasites in the blood, the child becomes more resistant to other common infections, e.g. broncho-pneumonia, and other side-effects occur and (2) an adequate control is impossible because people in recent years are becoming more 'health conscious' and use antimalarial drugs, insecticidal sprays and other powerful medicaments, few of which were available when the experiment began. The most obvious measurement is malarial mortality (including infantile mortality) rather than morbidity, but in countries where statistics are not kept and where the absence of facilities for autopsies entails ignorance of the cause of death, it is impossible to obtain figures of much significance. With other more spectacular diseases, like sleeping sickness, yellow fever, cerebrospinal meningitis or smallpox, diagnosis is simple and the ravages of these infections can be more precisely ascertained.

MALARIA

Malaria provides a good example of the change in pattern of a disease. Until a few years ago malaria was probably the greatest single killer of the human race and half the total mortality of man was claimed to be due to this infection. The discovery of the insecticidal properties of DDT in 1942 revolutionized the situation. Before that date, we struggled to control the disease with our cumbersome antimalarial measures of drainage, spraying with paris green or oil, and other more fanciful methods like zooprophylaxis, the introduction of larvivorous fish into mosquito breeding places, the automatic flushing syphon of George Macdonald, etc., etc.

In the Heath Clark lectures of 1934, Hackett spoke about malaria in Europe and felt that there was no justification for pessimism in the face of the malaria problem, and believed that eradication of the disease should not wait for better social organization and higher education, but should precede such improvements. All or most of Hackett's experience at that time was based on European malaria; today there is a strong tendency towards the idea that eradication must follow the introduction of properly organized medical and social services, and the explanation of the current failure of malaria eradication in Africa is thought to be because these services scarcely exist in many parts of the continent. Since Hackett's time, the perspective has changed and we are at last facing up to the reality of the 'human factor' which had too often been forgotten.[§]

The introduction of the new insecticides nevertheless brought about an utter change in outlook; the rather pedestrian concept of malaria control was replaced by the utopian ideal of global eradication, and in the Heath Clark Lectures of 1953, Paul Russell preached fanatically on 'Man's Mastery of Malaria'. Though he was careful to say that 'to master' does not imply an end to the matter, he was confident at that time that malaria was 'well on its way to oblivion'. His optimism proved justified in regard to Europe, North America, and in a few limited regions of the tropics. Success was nearly reached in India where there used to be a hundred million cases of the disease every year and now there are fewer than a hundred thousand, while in some countries of South America, e.g. Venezuela, malaria, by the energy of Gabaldon (1968), has been made to disappear from all parts except the remote forests, where sylvatic mosquitoes continue to transmit the infection in the unprotected out-of-doors. The general death rate has dropped precipitously in these places and the population has correspondingly risen, but the changed situation has resulted in problems almost as formidable as the disease itself. The vital statistics of Mauritius are often quoted: after the elimination of malaria from this island in 1948, the birth rate rose from 35 to 45, the infant mortality fell from 150 to 80 and the general mortality from 26 to 12 per 1000; as a consequence the population increased in 13 years by 50 per cent. As Burnet (1966) prophesied, Mauritius is now facing an explosive situation.

Even in a well-organized society like that of Ceylon, where eradication of malaria seemed to be just around the corner, the disease may return in a fulminating epidemic form; over 2 million new cases in fact occurred in that island in 1968 and 1969.

Elsewhere in South-east Asia and over much of Africa, the disease remains practically untouched by modern anti-malarial measures. This is due not only to the general lack of organized health services and of skilled personnel in these parts of the tropics, but, particularly in Africa, to the presence of what is probably the most dangerous malaria vector in the world: *A. gambiae*. We have seen the ravages which this mosquito can produce when it is introduced into new countries. The mosquito was transported in 1930 accidentally in a French ship from West Africa to the coast of Brazil; it started to breed in profusion and by 1938 it had occupied 12,000 sq. miles of country where violent epidemics of malaria broke out (Soper and Wilson, 1943). The exotic *A. gambiae* was a much more efficient vector than the local *A. darlingi*. Similarly, during the Second World War, *A. gambiae* crept insidiously into upper Egypt and the threatened epidemics were only suppressed by the prompt action of Madwah (of the Egyptian Government) and of Austen Kerr (of the Rockefeller Foundation) in a joint campaign (see Busvine, 1948).

Human malaria is thought to have originated in Africa many thousands of years ago as a zoonosis derived from gorillas and chimpanzees, which still harbour parasites identical or practically identical with the human ones today (see p. 49). The indigenous inhabitants of the continent have thus had ample opportunity of adaptation to these parasites and it is not surprising that in the course of evolution, they have inherited a considerable degree of tolerance. Over much of Africa, also, acquired immunity plays a greater part in controlling the severity of the disease in rural areas than do any efforts by the physician or hygienist.

The incidence of malaria is much lower in the large towns where control measures were until lately concentrated. This work needs highly qualified staff, but above all, single-minded devotion to the task: without the latter, the campaigns made little impression, for as Emerson wrote, 'nothing great was ever achieved without enthusiasm'. Manuwa (1968) in an important address to the Eighth International Congress of Malaria in Teheran, emphasized the need for mass campaigns in Africa, specialist controlled, with well-trained staff and ample financial resources, and integrated with the general educational, agricultural and public health programmes of the country. This is the materialistic approach, but without enthusiasm it is but a wasted effort.

The accelerated development which began after the First World War was accompanied, in East Africa, by a diffusion of malaria. The railways and roads transported *A. gambiae* and *A. funestus* from the malarious lowlands into the hills. At these higher altitudes, malaria, paradoxically, took a greater toll than in the lowlands, for neither natural nor acquired immunity existed in the indigenous population of the mountains. The immigrant races from Europe always suffered from the worst effects of malaria, and unless they took rigid precautions, they frequently died of pernicious forms of the disease or suffered from cachexia. The most dramatic syndrome was blackwater fever, due to sensitization by small doses of quinine of people infected with *Plasmodium falciparum*; once this drug had dropped out of the vade mecum, blackwater fever vanished and the condition has hardly been seen since 1945.

The tide actually began to turn when atebrin and plasmoquin were introduced in 1933; these powerful drugs were more effective than quinine both for cure and prophylaxis, and when chloroquine and primaquine arrived after the Second World War, an even greater effect became apparent. But the most significant alteration was produced by the introduction, about the same time, of DDT and other insecticides. A wave of optimism arose that country-wide eradication might be possible, and such a policy was widely appreciated after the 1st African Malaria Conference of WHO held in Kampala, in 1950.

The insecticides were applied in the towns throughout tropical Africa where the malaria statistics quickly showed great improvement. But though the figures fell also in rural areas where the houses were sprayed, e.g. in Kericho in Kenya, in the Taveta-Pare region of Tanzania and in Kigezi, Uganda, eradication was never complete. The deliberate eradication campaigns in Zanzibar and Pemba, based on WHO principles attained a higher degree of success, in that the incidence of malaria sank in 1967 to 2.6 per cent and 0.1 per cent respectively, but actually failed in that final eradication was never achieved. A similar result was reported in two small territories in West Africa. But a surprise was in store which perhaps vitiates the value of the above figures as an indication of the effects of control: the malaria statistics *all over* Africa began to decline *irrespective* of deliberate anti-malarial measures. Those populations in the low country of Nyanza in Kenya which thirty years earlier had shown 75 per cent parasite rates, now show under 50 per cent. In similar country on the Uganda side of the frontier, Onori and Bentheim (1968) have reported equally low figures today, i.e. low in comparison with the earlier ones, and the parasite rate in the 5–15 years group was on an average only 40 per cent. It was suggested that milder malaria years were responsible for this remarkable drop in incidence, but the effect persisted and even years of high rainfall were unaccompanied by any severe recrudescence of the disease. A likely explanation seems to be that the people themselves, with better education, have become more aware of the nature of malaria and have taken advantage of the liberal stocks of the potent new drugs and DDT sprays now available in many stores throughout Africa.

Any gain in the balance-sheet of malaria as the result of a decline in severity of the disease has been more than offset by the development of resistance by the parasite to drugs, a phenomenon which Sergent designated 'mithridatism' after Mithridates IV, ruler of the kingdom of Pontus (the modern Armenia) who rendered himself immune to poisons by taking gradually increasing doses, and thus frustrated the attempts of his enemies to get a new King.

Malariologists watched with apprehension the spread of resistance in the tropics, and reports from uncritical observers suggested that the situation was worse than it really was. The present position has been critically examined by Schnitzer and

Hawking (1963) in a review of the whole subject of drug resistance, and in special reference to malaria by Peters (1969); and to the strategy of malaria eradication by Bruce-Chwatt (1968), Busvine (1969) and by a group in WHO (1968). The emergence of drug-resistant strains on a wide scale could clearly have a devastating effect on the progress of such campaigns.

The seriousness of the problem was first appreciated in 1948 by the failure of proguanil to act as a prophylactic against malaria; 'break-throughs' were reported in Malaysia (Edeson, 1950) and in East and West Africa, but the phenomenon was limited to a few places in these countries. Soon afterwards, the other drug commonly employed in chemoprophylaxis—pyrimethamine—was noticed to be losing its efficacy, first in Makueni in Central Kenya (Avery-Jones, 1952) and later in Tanzania where Clyde (1954) traced in great detail the slow dispersal of resistant strains beyond the periphery of the treated zone: an extension of 20 miles in 5 years and of 100 miles in 10 years. It should perhaps be noted that the populations on which these observations were made, were suffering from holoendemic malaria and frequency of transmission plus the high immunity may make this a 'special case'.

But much more important than the development of resistant strains of malaria parasites to these two drugs was the appearance of strains of P. falciparum resistant to chloroquine, the 4-aminoquinoline which had come to supersede all others in the ordinary treatment of cases of malignant tertian malaria. The first reports came from the Magdalena Valley in Colombia in 1959. The writer was in that valley in the autumn of that year and heard many rumours about resistance, especially in the forests on the Colombian-Venezuelan border, where the inability of insecticides to reach the sylvatic vector, A. nunez-tovari was hindering the final eradication of malaria from Venezuela. Chloroquine resistance has now been confirmed in many parts of South America.

The warfare in South-east Asia in recent years has been accompanied by the emergence of chloroquine-resistant strains of P. falciparum in that region, particularly in Malaysia and Vietnam, but also in Thailand and Cambodia. The danger lies in the fact that patients infected with this strain of P. falciparum fail to respond to most other drugs also—there is cross resistance to the other 4-amino-quinolines, to proguanil and pyrimethamine, mepacrin and even quinine.

An example of cross resistance to antimalarial drugs was recently reported by Killett *et al.* (1968), which illustrated not only this phenomenon but a number of other interesting features. The case was a British soldier who had been stationed in North Perak, where he had first been on proguanil and then on chloroquine for prophylaxis against malaria. In spite of this treatment which he conscientiously followed, he developed fever and was evacuated to England. He was treated at the Queen Alexandra Military Hospital with 3.1 g chloroquine, but although parasites disappeared, he continued to have low fever, anaemia and splenomegaly. Twenty days later a recrudescence occurred and malaria parasites reappeared in the blood; this time he was given quinine for seven days, but on day 46 another attack came on, which was accompanied by a low parasitaemia. On this occasion, he was treated with pyrimethamine, potentiated with sulphor-methoxine and the falciparum infection was finally eliminated. Nevertheless, on day 69, the patient once more returned to hospital, now with an attack of vivax malaria. The sporozoites of *P. vivax* had been held in latency by the concomitant infection of *P. falciparum*, a well-known but completely inexplicable phenomenon which is further discussed on p. 10. *P. vivax* never becomes drug-resistant in the ordinary sense, and the infection in this case, once aroused from dormancy, immediately responded to chloroquine. Nevertheless it may be noted that the exoerythrocytic (post-sporozoite) stages of this strain of *P. vivax* were unaffected, i.e. they were resistant to the prolonged exposure to drugs, including proguanil and chloroquine in Malaya and quinine and a pyrimethamine-sulphone mixture in England.

The writer had the privilege of seeing the above case, and of taking his blood, infected with *P. falciparum*, for culture (Bass and Johns). On examination of the parasites 24 hours later, it was most surprising to find 'exflagellating crescents'. Exflagellation normally only takes place for 10–20 minutes after the blood has left the circulation—in the mosquito's gut or *in vitro* on a slide. Here the process had been delayed for 24 hours; it is true

TABLE 1. Resistance of anopheline vectors to insecticides. (From Busvine and Pal, 1969 and Hamon and Pal, 1968.)

Place	Vector	Insecticide	Level of Resistance	Comments
Greece	sacharovi	DDT Dieldrin	Serious	First seen 1951
Turkey	sacharovi	DDT Dieldrin	Serious	
Portugal	maculipennis complex	DDT	Low	
Israel	sacharovi	DDT	Serious	
U.A.R.	pharoensis	DDT Dieldrin	Serious	
Persian Gulf	stephensi	DDT Dieldrin	Serious	
Iraq	stephensi	DDT Dieldrin	Serious	First seen 1957
Pakistan	stephensi and culicifacies	DDT Dieldrin	Widespread	
India	stephensi	DDT and HCH	Widespread	
	culicifacies	DDT	Widespread	HCH susceptible

Location	Species	Insecticide	Severity	DDT status
Brunei	balabacencis	DDT		
Philippines	flavirostris	Dieldrin	Moderate	
U.S.A.	quadrimaculatus	DDT Dieldrin	Serious	DDT susceptible
Mexico	pseudopuncti-pennis	Dieldrin	Serious	in places
Central America	albimanus	DDT	Serious	in places
Venezuela	albimanus aquasalis and albitarsis	Dieldrin (DDT)	Serious	
Colombia	albimanus albitarsis	Dieldrin		DDT susceptible
	albitarsis	Dieldrin DDT	Serious	DDT susceptible
Peru	pseudopuncti-pennis	Dieldrin		DDT susceptible
Senegal	gambiae	DDT		DDT susceptible
Upper Volta	gambiae	DDT		DDT susceptible
Tropical Africa (parts)	gambiae	HCH and Dieldrin		Incipient DDT resistance
Madagascar	gambiae	Dieldrin		
West Africa (parts)	funestus	HCH Dieldrin		

that exflagellation can be halted by preservation of the infected blood at $-77°C$ and be resumed after thawing, but at the temperature of $37°C$ at which the culture was kept, the phenomenon was most unexpected. Bishop and McConnachie (1956) showed that some unknown substance in the *plasma* is essential for exflagellation, but beyond this knowledge and the fact that the process is inhibited by lowering the pH of the medium below 7.2, we are entirely ignorant of the biochemical factors concerned in this essential part of the life cycle of malaria parasites.

Both resistance of parasites to drugs and of the anopheline vectors to insecticides could have a profound effect on the progress of malaria eradication. At present, the latter seems to be the greater danger and the global situation is summarized by Busvine and Pal (1968) and is shown in Table 1. Nevertheless, these drawbacks have not influenced the major campaigns to any great extent, and the real hold-up is the lack of organization and resources in the tropical countries where malaria still prevails.

ONCHOCERCIASIS

Onchocerciasis is a purely tropical disease which the Director-General of WHO recently stated is likely to spread far afield as the result of man's activities, such as the installation of the Volta Dam. The disease has travelled in the recent and in the more remote past, and in doing so its pattern has changed. I want therefore to discuss this subject, particularly as I have had the opportunity of studying onchocerciasis in most parts of the world where it occurs: in Kenya, Uganda and Tanzania, in the Congo, Cameroun, Nigeria and Ghana, and in Mexico and Guatemala.

I missed discovering onchocerciasis in Colombia by a short distance. I was in South America in 1959, and in discussions with my host, Carlos San-Martin, I asked him if there were any large African settlements in Colombia where fast-running rivers were a feature of the locality. West African immigrants might be supposed to have introduced the infection, and such rivers were likely to breed *Simulium*, the vector of the filarial worm *Onchocerca volvulus*. He replied that precisely such an environment existed in the Choco—a rather remote region on the Pacific slopes of the western Cordilleras. I went by 'plane to this place with Rafael Samper, the well-known surgeon of Bogotá, who promised to help me in investigating the population. In spite of torrential rain, we went in speed boats down the San Juan river and then up the Tamaná; it was a dramatic passage over the turbulent waters with whirlpools. After many hours, we arrived at the African village of Tigre, and set about examining the people—Samper doing the skin-snips and I examining them with a Macarthur microscope for microfilariae. It seemed an ideal place, with plenty of *Simulium*, 90 per cent humidity and a countryside reminiscent of the Cameroons, but all the preparations were negative, including others that we examined in similar villages nearby. A few years later, Assis-Masri and Little (1965) diagnosed a case of onchocerciasis in a Negro, who lived further south, where the incidence of the disease is now being intensively studied (43 people in 312 examined were found to be infected).

Onchocerciasis probably reached the New World in African slaves at the time of the major exportations in the eighteenth century or perhaps earlier. Transmission was effected by American vectors, particularly *Simulium ochraceum*, which has different habits from the African species, and in the course of years the filarial worm is thought to have slightly changed its nature. It seems to give rise more often to ocular lesions, and Robles' disease, as it is called in central America, was recognized from the time of its discovery (1915) in Guatemala to be characterized by blindness. Brumpt thought that a different species of helminth was concerned and named it *O. caecutiens* (= the blinding *Onchocerca*).

Duke *et al.* (1967) in the last few years have shown the biological distinction between the Old and New World species, by taking infected Africans to Guatemala and demonstrating that their microfilariae will scarcely develop in the American species of *Simulium*. On the other hand, when the Guatemalan strain of *O. volvulus* was passed into a chimpanzee, the African *Simulium damnosum* practically failed to become infected when this species fed on it. What Duke has not yet done, is to try to adapt the respective strains to the exotic vector: the difference in response might be because two species of *Onchocerca* are concerned, one in America and the other in Africa. Although onchocerciasis is particularly prevalent in African 'pockets' in Colombia and Venezuela, the disease occurs further north in places like western Guatemala and Chiapas in Mexico, where the African has scarcely penetrated. On the contrary, when David Lewis and I were in British Honduras in 1959, we made a point of examining the Africans who are numerous in Stann Creek and where the local *Simulium metallicum* is a terrible pest; yet these Africans were found to be uninfected. We fear that in time onchocerciasis is bound to invade other territories in Central and South America in places where suitable vectors occur.

The disease has already shown signs of spreading in tropical Africa. Presumably onchocerciasis has existed in that continent since remote times, and it is likely that in the course of evolution of the primates, the infection was first present in chimpanzees, for these animals are easily infected with *O. volvulus* (Duke, 1962) and that a zoonosis involving primitive man occurred.

Our actual knowledge of onchocerciasis dates from recent times. Cases were first recognized by missionaries in the Gold Coast in 1893, and specimens of the worm were sent to Leuckart who gave the parasite its name. The vector of *O. volvulus*, *Simulium*, was discovered by Blacklock in West Africa in 1926. My first encounter with this insect was in the same year, when my wife and I were badly bitten by hordes of these flies when we bathed in the Ruimi River on our way up Ruwenzori to climb Mount Baker; yet, although I had just done the course of Tropical Medicine at the London School, Blacklock's discovery of the vector in Sierra Leone had not then been ventilated and so I took no particular notice of these insects.

Dr Geoffrey Timms was one of the first to recognize the infection in Kenya in 1937. The disease and *S. damnosum* were found in many parts of Uganda by Gibbins (1956) and was later described in great detail by Barnley (1949) and Nelson (1958). Onchocerciasis was discovered in Tanzania in 1957 under rather curious circumstances. Professor Mario Giaquinto was giving a course of WHO lectures on onchocerciasis and malaria at the Malaria Institute of Amani. He wanted to demonstrate the skin-snip technique to the students and took skin-snips from local inhabitants. To everyone's surprise, he found that the biopsies were positive and he was able to demonstrate not only the technique but actual microfilariae to his class. Since then the disease has been found much further south in Tanzania, to which it has apparently spread in recent years.

A growth in the population living in the fertile areas where onchocerciasis is common is followed by increased transmission of the infection until, as happened in Kodera in Kenya, practically everyone eventually contracts the disease. This no doubt is happening over much of tropical Africa, and it is highly desirable that these fertile lands should be made fit for occupation.

Fortunately a method exists, for in 1946 McMahon and I discovered a technique which was to eradicate entirely the vector species (*S. neavei*) from the endemic areas of Kenya. It was a simple enough procedure—dripping DDT emulsion into infested rivers and streams after measuring in 'cusecs' the rate of river flow. The difficulty lay in mapping these rivers, and McMahon in the course of his work in Kenya must have walked thousands of miles. By 1955, the whole of the endemic regions of Kenya had been subjected to this treatment and transmission of the infection had ceased. Not a single *S. neavei* has been caught since that date (McMahon, 1967). Roberts *et al.* (1966) gave a final report of the situation, based on extensive surveys done 9, 11 and 18 years after the original application of DDT. No sign of the disease was found in anyone born after interruption of transmission; adult worms in nodules and microfilariae in the skin were detected in persons examined 11 years later, but none in those observed 18 years after transmission had ceased. From these figures, it was concluded that the adult *O. volvulus* can live as long as 11 years, but dies before 18 years. Earlier work in the Kaimosi Forest of Kenya had indicated that the 'incubation period' (the interval between the infected bite and the appearance of microfilariae in the skin) of onchocerciasis could be as short as 5 months. It is sad to think that new cases of onchocercal blindness are still appearing in the inhabitants of these regions long after the vector had been eradicated.

Other vectors of onchocerciasis are proving more difficult to eradicate, though their numbers have been greatly reduced. Barnley (1949) applied a similar technique to the Nile at Jinja and temporarily abolished *S. damnosum*, while Davis reduced the density of this fly by 95 per cent over an area of 2000 sq miles in Nigeria, where the infection could no longer be found after five years in the younger children. Wanson *et al.* (1950) successfully rid Léopoldville of *S. damnosum* by spraying DDT from helicopters and only now has the fly returned to this city after nearly twenty years of freedom.

Woodruff *et al.* (1966) stress how the pattern of onchocerciasis varies in the savannah and forest regions of Africa; the former has much blindness in the population and the latter, little. But the explanation of the difference still eludes us, while the controversy over the onchocercal aetiology of lesions in the posterior segment of the eye also remains unresolved.

THE CHANGING PATTERN OF PARASITIC DISEASES ON THE COURSE OF THE KUJA RIVER IN KENYA

The two previous examples have been drawn from diseases which have been notably influenced by control measures. It is essential to bear in mind how natural factors may affect the incidence of disease; too often we ascribe our small successes in improving the public health to our own efforts, while in reality it was a 'kind season' which was perhaps responsible. Geography and climate can have a profound effect on the disease pattern. A study of the influence of altitude on natural foci of infections in a limited area may therefore be rewarding and a brief description follows of the situation along the course of the Kuja River in Kenya, illustrating these points (Garnham, 1965). The source of this river lies in the Kisii Highlands at an altitude of about 2100 m; the Kuja then descends westwards in cascades into a broad valley where the flow is lessened for many miles at an altitude of around 1700 m. (Plate III, Fig. 11.) The Kuja is joined by a large tributary, the Riana river, 300 m below this point, and by the Gori near where the main river empties its waters into Victoria Nyanza on the Tanzanian frontier, just south of the Equator, and at an altitude of 1100 m. During the thousand metres of descent of the river, and over a length of 200 km, there is a great change in the natural features;

and these are reflected in the natural foci of infections. The mean temperature at lake level is 24°C and at the source 18°C; this difference corresponds to the well-known relationship of half a degree C for every 100 m.

The highlands around the source of the river are prone to severe epidemic malaria, probably entirely man-made in origin. Until the country became developed, the malaria vectors (*Anopheles gambiae* and *A. funestus*) were absent. Motor and rail transport after the First World War brought up these mosquitoes from low-lying areas around Victoria Nyanza. The introduction of the ox wagon entailed the construction of rough roads, in which wheel ruts were common and near which borrow pits inevitable; these provided ideal breeding places for *A. gambiae*, while the attempted drainage of natural swamps provided facilities for *A. funestus*. The development of the country meant also that people from malarious places drifted into the area, and these provided excellent reservoirs of infection. Epidemics inevitably followed when the climatic conditions became suitable, which was not necessarily every year. The trigger which sets off an epidemic has never been exactly identified.

The Gori river is a large tributary of the Kuja which rises not far from the source of the main river itself, and at an altitude of 1800 m runs through typical game country, where Rhodesian sleeping sickness occurs (cf. Fig. 7). The natural focus of this infection is constituted by the following 5 elements:

1. Acacia thorn bush and evergreen thickets, extending a considerable distance from the river.

2. Two species of tsetse fly—*G. swynnertoni* and *G. pallidipes*.

3. Game animals upon which the tsetses feed, and in particular the bushbuck (*Tragelaphus scriptus*).

4. *Trypanosoma rhodesiense*, the organism responsible for the acute form of sleeping sickness in man.

5. Man. Sporadic cases of zoonotic origin occur but epidemics do not break out as the human population is too small.

During and after the Second World War, it is probable that such infected persons crossed the lake by dhow to Uganda and introduced *T. rhodesiense* into the densely populated district of

Busoga. Terrible epidemics arose and spread back into Kenya, where recently the infection has returned to tributaries on the north banks of the Kuja, and into the Lambwe Valley. Glover (1962) reported 17 cases here, near the site of the natural focus, where *G. pallidipes* is again the vector, and the numbers have now increased to a hundred *per annum*. Owing to the steepness of the hills around the Lambwe, insecticide spraying from the air is impracticable, and it is said that this species of tsetse can only be eliminated by discriminative bush clearing or defoliation of the trees. The measure would destroy the essential habitat of Jackson's hartebeeste and the roan antelope, and cause the extinction of the former animals from Kenya. This 'parasitologist's dilemma' could be resolved by moving the small human population and leaving the valley to its natural inhabitants (game and tsetses) or by applying the insecticide by special methods from the land.

We may leave the Kuja River for the moment, in order to observe the further events which followed the introduction of *T. rhodesiense* into Uganda and on the Uganda/Kenya border. Here, instead of remaining confined to the vicinity of the river or lake, a new vector, *Glossina fuscipes* suddenly extended its breeding habits and invaded the hedges round the native huts and villages (Plate III, Fig. 12). These hedges were originally composed of euphorbia bushes, but in the last decade, the ornamental *Lantana camana* has been extensively planted and has overgrown the original vegetation (van den Berghe, 1965): a new niche for the infection was thus created on the door-step of human habitations. The fly began to transmit *T. rhodesiense*, not only to the people, but to the cattle, introducing a new zoonotic element into the situation, though Onyango *et al.* (1966) discount the importance of these animals as permanent reservoirs.

Let us resume our trip down the Kuja river and see how its tributary—the Riana—provides suitable conditions at 1700 m for another infection, onchocerciasis. The features of the natural focus are as follows:

1. A fast running river and tributaries with well-wooded banks, and a hilly terrain, and below an altitude of about 2000 m (cf. Fig. 11).

2. The presence of freshwater crabs (*Potomonaustes niloticus*).

3. The vector fly, *Simulium neavei*, whose immature stages are found on the crabs in cascades in the river.

4. An extinct factor: the chimpanzee, the probable reservoir of the worm, *Onchocerca volvulus*, in this region, in the past. The chimpanzee probably lived here not very long ago, for in neighbouring forests, its presence is still remembered in legends of the 'Nandi Bear'.

5. The nematode worm, *Onchocerca volvulus*.

6. The human population living near the river. Unlike *S. damnosum*, the other important vector of African onchocerciasis, *S. neavei* has a short flight range and the natural focus is therefore limited to a kilometre or less from the river banks.

Human onchocerciasis became widespread in the Kuja Basin, and at Bassi, Buckley (1959) noted that a third of the population was infected. The infection spread, or was coexistent, further downstream at Riana and in the adjoining river systems where the disease was so severe in its ocular manifestations that the place was called the 'Country of the Blind'.

The natural focus of this disease can be eliminated by the removal of any one of the above five factors, and Buckley (1961) accomplished this in an elegant experiment by changing the character of the riverine bush: discriminative clearing of under-growth and the smaller trees was enough to destroy the habitat of the vector.

Below the onchocercal country, the hills come down to the plains and where these two configurations meet, new natural foci of infections are born, for it is particularly at such junctures that the 'edge effect' encourages tick typhus, arbovirus infections, and here on the Kuja, plague and Gambian sleeping-sickness.

Plague used to be a feature of this region, but for some unknown reason, the infection disappeared 30 years ago and has not returned. The last outbreak was in 1938 in villages near the confluence of the Nyangueta and the Kuja. The mud and wattle huts with earth floors, and grass huts, still provide ideal conditions for rat infestation, but the disease is inactive, though according to Davis *et al.* (1968) in the provincial capital

(Kisumu) 12 per cent of the rats are still serologically positive for *P. pestis*. Moreover, in the neighbouring province, Heisch *et al.* (1953) have shown the existence of sylvatic plague; this was in veldt country with a much lower rainfall, very different from the lush countryside around the Kuja, where if a wild rodent reservoir of plague exists, the species involved are unlikely to be the steppe rodents which were found infected further east. It is however possible that the isolated epidemics of plague which used to break out along the Kuja and elsewhere in Kavirondo country, arose from the importation of plague-infected merchandise from Uganda.

There are conflicting theories (see Mulligan, 1970) about the introduction of Gambian sleeping sickness into East Africa, but it is probable that the disease invaded the shores and islands (Plate IV, Fig. 13) of Lake Victoria a hundred years ago, in its furthest spread eastwards from the Congo at the time of the Stanley expedition (Morris, 1960). The infection has persisted ever since, as a West African relic, like certain birds, reptiles, and mosquitoes, and has continued to devastate the country in a series of alternating endemics and epidemics. The Kuja river forms a natural focus of sleeping sickness of this type from the Kabwoch forest to the lake and certain well-defined elements are concerned, as follows:

1. Riverine forest between an altitude of 1300 and 1100 m. The Kuja flows slowly through this stretch, the vegetation on the banks expanding and contracting in width, but the habitat of the vector fly, as Buxton (1955) states, is the narrow strip along the edge of the water.

2. The presence of *Glossina fuscipes*.

3. A varied array of animals has been suspected of acting as reservoirs, including the sitatunga, guinea fowl and crocodile. Such animals have been inoculated with the trypanosome, and an infection of longer or shorter duration has followed. The buck is the most likely reservoir in this group, because artificially infected animals have proved capable of infecting tsetse flies. The crocodile on the other hand shows only fleeting trypanosomes after inoculation with infected blood, and flies fed on these animals failed to become infected (Willett, 1960). Yet the

preferred host of *G. fuscipes* is the crocodile, and its role as a reservoir cannot be entirely dismissed. Domestic animals, such as pigs and goats, were considered by van Hoof (1947) as possibly dangerous reservoirs, but no direct evidence has yet been obtained.

4. *Trypanosoma gambiense*, the organism responsible for chronic sleeping sickness, is well adapted to man in that the disease usually takes several years to kill the host.

5. The human population living within range of *G. fuscipes*. Only fisherman or people visiting the river contract trypanosomiasis and the first question that a suspected case is asked is, 'Where do you live?' for if the reply denotes residence away from the river, a diagnosis of this form of sleeping sickness can be dismissed.

Sleeping sickness along this lowest part of the Kuja river is interesting, as providing an example of a natural focus of potential disease where the vector presumably has always been abundant, but where until the beginning of this century the parasite was absent, either in man or animals. Once introduced, in the form of a carrier, the focus quickly became established. The active natural focus can be eliminated in various ways, of which several have been tried. The high riverine vegetation can be removed and the habitat of the vector is ruined; the tsetse may be hand-caught or killed by DDT spraying; the human population can be removed from the fatal riverine zone, or given large enough clearings to avoid contact with the fly. All these measures, and others such as destruction of the trypanosome by drugs, have been used. The literature pertaining to tsetse flies is full of ingenious attempts to get rid of this pest, but all pale into insignificance compared to the use of synthetic insecticides by which Glover *et al.* (1958) have virtually exterminated the fly from the Kuja basin and lake shore.

Wilcocks (1962) in his Heath Clark lectures on sleeping sickness in East Africa is careful to point out that no permanent freedom from the disease can be anticipated unless the population is convinced of the need for a new way of life and is willing to accept progress in all directions.

General infections of all kinds abound on the lake fringe near the outfall of the river; a good example is *Schistosoma mansoni* which is common in Kadem. This parasite produces a form of bilharziasis which still retains much of the features of a natural focus, unlike *S. haematobium* which easily loses its focal epidemiology. *S. mansoni* is eclectic up to a point, but the following conditions tend to concentrate and augment the infection:

1. Slowly running water, of a permanent nature, such as occurs at the mouth of the river and the lake shore nearby.
2. Snail hosts: *Biomphalaria pfeifferi* and *B. sudanica*.
3. An animal reservoir. The infection originally prevailed in wild animals hosts, and even today in some parts of Africa, these may play an important role in transmission. Secondary domestic hosts, such as dogs, can be involved. Nelson *et al.* (1962) found two infected dogs near Lake Victoria, and more than 50 per cent of baboons have been discovered to harbour *S. mansoni* in endemic areas. The most recent evidence incriminating baboons as a reservoir of human schistosomiasis comes from Lake Manyara National Park in Northern Tanzania. A research station was established in a remote corner of the Park in 1966, and below the camp lie a series of rock pools, in which visitors used to bathe; 15 out of 17 visitors, according to Fenwick (1969) developed severe 'swimmer's itch'. They became ill 23 days later with fever (up to 41°C), wasting, cough, and a bloody diarrhoea. *S. mansoni* ova were found by the 45th day after exposure. A large troop of baboons (*Papio anubis*) frequented this site, and their excreta were deposited all around. Ova were easily found in this material, and adult schistosomes were seen in 7 baboons which had been shot and dissected. The infection was still active in these animals, 18 months after swimming in the pools had been stopped, and it is clear from these data that the baboon can be more than a mere accidental host of the parasite.

Widespread infection with *Schistosoma mansoni* exists in other monkeys also; Cheever *et al.* (1970) found eggs in 11 out of 47 *Cercopithecus aethiops* caught near Musoma on Lake Victoria and in 6 of 18 from Asmara in Somalia. Some of these animals showed granulomatous changes in the liver, colon and lungs.

4. Around the mouth of the Kuja river, the people contract

bilharziasis by washing and bathing in the edges of the river or lake, where the cercariae penetrate the unbroken skin. High rates of infection occur in this area, though the Luo population apparently possesses a degree of natural resistance.

The natural foci on the Kuja river are all dependent on altitude; epidemic malaria ceases below the 2000 m contour but becomes holoendemic at and just above the lake level, and it is in such places that blackwater fever used to be frequent; onchocerciasis has a specific niche around the cascading waters below 1800 m, and above 1300 m; the natural home of Rhodesian sleeping sickness is at about 1800 m but with changed conditions it descends to 1200 m, while Gambian sleeping sickness is limited by the distribution of *Glossina fuscipes* which does not extend much above 1300 m; Mansonian schistosomiasis is confined on the Kuja (though not elsewhere) to the country near the lake shore.

Although the true natural foci of infection remain fixed, they may become altered by man's activities, which may either accidentally or deliberately abolish them altogether. Sleeping sickness, particularly, is a condition which disappears fortuitously with the advance of civilization, or designedly after the destruction of tsetse flies. As in all aspects of life today, parasitic diseases are subject to change at an alarming pace and in all directions; new ecological situations are arising, and the 'White Man's Grave' is more likely to be found now in New York or London than in the Bay of Guinea or Bengal.

REFERENCES

Assis-Masri, G. and Little, M. D. (1965) 'A case of ocular onchocerciasis in Colombia', *Trans. R. Soc. trop. Med. Hyg.* **59**, 717.

Avery-Jones, S. (1954) 'Resistance of *P. falciparum* and *P. malariae* to pyrimethamine following mass treatment with this drug', *E. Afr. med. J.* **31**, 47–49.

Barnley, G. R. (1949) 'Onchocerciasis in Kigezi district, Uganda', *E. Afr. med. J.* **26**, 308–10.

Biagi, F. F., Biagi, A. M. and Beltran, H. F. (1966) '*Phlebotomus flaviscutellatus*, transmissor natural de *Leishmania mexicana*', *Prensa med. Mex.* **30**, 267–72.

BISHOP, A. and MCCONNACHIE, E. (1956) 'A study of the factors affecting the emergence of the gametocytes of *Plasmodium gallinaceum* from the erythrocytes and the exflagellation of the male gametocytes', *Parasitology*, **46**, 192–215.

BLACKLOCK, D. B. (1926) 'The development of *Onchocerca volvulus* in *Simulium damnosum*', *Ann. trop. Med. Parasit.* **20**, 1–48.

BRUCE-CHWATT, L. (1968) 'Drug resistance of malaria parasites and malaria eradication', *8th Internat. Congr. trop. Med. Malaria* 1, 1398.

BUCKLEY, J. J. C. (1949) 'Studies on human onchocerciasis and *Simulium* in Nyanza Province, Kenya. I. Distribution and incidence of *O. volvulus.*', *J. Helm.* **23**, 1–24.

BUCKLEY, J. J. C. (1951) 'Studies on human onchocerciasis and *Simulium* in Nyanza Province, Kenya. II. The disappearance of *S. neavei* from a bush-cleared focus', *J. Helm.* **25**, 213–22.

BURNET, M. (1966) *Natural History of Infectious Disease*, Cambridge: University Press.

BUSVINE, J. R. (1969) 'The impact on malaria of insecticide resistance in anopheline mosquitoes', *Trans. R. Soc. trop. Med. Hyg.* **63**, Suppl. 19–30.

BUSVINE, J. R. and PAL, R. (1969) 'Impact of insecticide-resistance on control of vectors and vector-borne diseases', *Bull. Wld Hlth Org.* **40**, 731–44.

CHEEVER, A. W., KIRSCHSTEIN, R. L. and REARDON, L. V. (1970) '*Schistosoma mansoni* infection of presumed natural origin in *Cercopithecus* monkeys from Tanzania and Eth opia', *Bull. Wld Hlth Org.* In Press.

CLYDE, D. and SHUTE, G. T. (1959) 'Survival of pyrimethamine-resistant *Plasmodium falciparum*', *Trans. R. Soc. trop. Med. Hyg.* **53**, 170–2.

DOROLLE, P. (1968) 'Old plagues in the jet age. International aspects of present and future control of communicable disease', *Br. med. J.* **4**, 789–92.

DRAPER, K. C. and DRAPER, C. C. (1960) 'Observations on the growth of African infants with special reference to the effects of malaria control', *J. trop. Med. Hyg.* **63**, 165–71.

DUKE, B. O. L. (1962) 'Experimental transmission of *Onchocerca volvulus* from man to chimpanzee', *Trans. R. Soc. trop. Med. Hyg.* **56**, 271–80.

DUKE, B. O. L., MOORE, P. J. and DE LEON, J. R. (1967) 'Onchocerca-Simulium complexes. v.', *Ann. trop. Med. Parasit.* **61**, 332–7.

EDESON, J. F. B. and FIELD, J. W. (1950) 'Proguanil-resistant falciparum malaria in Malaya', *Br. med. J.* **i**, 147–51.

ELSDON-DEW, R. (1949) 'Endemic fulminating amoebic dysentery', *Am. J. trop. Med.* **29**, 337–40.

FENWICK, A. (1969) 'Baboons as reservoir hosts of *Schistosoma mansoni*', *Trans. R. Soc. trop. Med. Hyg.* **63**, 557–67.

FORRESTER, A. T. T., NELSON, G. S. and SANDER, G. (1961) 'The first record of an outbreak of trichinosis in Africa south of the Sahara', *Trans. R. Soc. trop. Med. Hyg.* **55**, 503–13.

GABALDON, A. (1968) 'Duration of attack measures in a malaria eradication programme', *Am. J. trop. Med.* **17**, 1–12.

GARNHAM, P. C. C. (1965) 'The transformation of natural into unnatural foci as seen in the course of a tropical river' in *Theoretical Questions of Natural Foci of Disease*, pp. 267–74, Prague: Acad. Sci.

GARNHAM, P. C. C. (1968) 'The changing pattern of disease in East Africa', *E. Afr. med. J.* **45**, 641–50.

GARNHAM, P. C. C. and McMAHON, J. P. (1947) 'The eradication of *Simulium neavei* Roubaud, from an onchocerciasis area in Kenya Colony', *Bull. Ent. Res.* **37**, 619–28.

GEIGY, R. (1968) in *Infectious blood diseases of man and animals*, II, Ch. 18, 'Relapsing Fevers', New York: Academic Press.

GIBBINS, E. G. (1936) 'Uganda Simuliidae', *Trans. R. Soc. trop. Med. Hyg.* **31**, 217–38.

GLOVER, P. E. (1962) 'The importance of ecological study in the control of tsetse flies', *Bull. Wld Hlth Org.* **37**, 581–614.

GLOVER, P. E., LEROUX, J. G. and PARKER, D. F. (1958) 'The extermination of *Glossina palpalis* on the Kuja Migori river systems with the use of insecticides', *I.S.C.T.R. 7th Meeting* Brussels 331.

HACKETT, L. W. (1937) *Malaria in Europe* (Heath Clark Lectures, 1934), London: Oxford University Press.

HAMON, J. and PAL, R. (1968) 'Practical Implications of insecticide resistance in arthropods of medical and veterinary importance', WHO/VBC/68. 106. (Not published.)

HEISCH, R. B., GRAINGER, W. E. and D'SOUZA, J. ST A. M. (1953) 'Results of a plague investigation in Kenya', *Trans. R. Soc. trop. Med. Hyg.* **47**, 503–21.

HOEPPLI, R. (1952) 'Early references to the occurrence of *Tunga penetrans* in tropical Africa', *Acta trop.* **20**, 143–53.

KELLETT, R. J., GOWAN, G. O. and PARRY, E. S. (1968) 'Chloroquin-resistant *Plasmodium falciparum* malaria in the United Kingdom', *Lancet* **ii**, 946–8.

MAEGRAITH, B. G. (1969) 'Importation of disease by travellers and immigrants', *Proc. 8th Internat. Congr. trop. Med. Malaria* Teheran, 1968.

MANUWA, S. (1968) 'Mass campaign as an instrument of endemic disease control in developing countries' *Br. med. J.* **4**, 634–8.

McMAHON, J. P. (1967) 'A review of the control of *Simulium* vectors of onchocerciasis', *Bull. Wld Hlth Org.* **37**, 415–30.

MORRIS, K. R. S. (1960) 'Studies on the epidemiology of sleeping sickness in East Africa. II. Sleeping sickness in Kenya', *Trans. R. soc. trop. Med. Hyg.* **54**, 71–86.

MULLIGAN, H. (1970) Ed. *The African Trypanosomiases*, with 27 contributors, London: George Allen & Unwin.

NELSON, G. S. (1958) 'Onchocerciasis in the West Nile district of Uganda', *Trans. R. Soc. trop. Med. Hyg.* **52**, 368–76.

NELSON, G. S. (1962) 'The role of animals as reservoirs of bilharziasis in Africa', *CIBA Foundation Symposium on Bilharziasis*, 127–49.

ONORI, E. (1967) 'Distribution of *Plasmodium ovale* in the eastern, western and northern regions of Uganda' *Bull. Wld Hlth Org.* **37**, 665–8.

ONYANGO, R. J., VAN HOEVE, K. and RAADT, P. (1966) 'The epidemiology of *Trypanosoma rhodesiense* sleeping sickness in Alego location, Central Nyanza, Kenya', *Trans. R. Soc. trop. Med. Hyg.* **60**, 175–82.

ORMEROD, W. E. (1961) 'The epidemic spread of Rhodesian sleeping sickness', *Trans. R. Soc. trop. Med. Hyg.* **55**, 525–38.

PATERSON, A. R. (1934/5) *Book of Civilisation*, Parts 1 and 2, London: Longmans, Green & Co.

PETERS, W. (1969) Drug resistance in malaria—a perspective', *Trans. R. Soc. trop. Med. Hyg.* **63**, 1–45.

ROBERTS, J. M. D., NEUMANN, E., GÖCKEL, C. W. and HIGHTON, R. B., (1966) 'Onchocerciasis in Kenya, Nine, 11 and 19 years after elimination of the vector', 3WHO/Oncho/66.55.

RUSSELL, P. F. (1955) *Man's Mastery of Malaria* (Heath Clark Lectures, 1953), London: Oxford University Press.

SCHNITZNER, R. J. and HAWKING, F. (1963) *Experimental Chemotherapy*, vol. 1. New York: Academic Press.

SOPER, F. L. and WILSON, D. B. (1943) *Anopheles gambiae in Brazil*, New York: Rockefeller Foundation.

SWYNNERTON, C. F. M. (1936) 'The tsetse flies of East Africa', *Trans. R. entom. Soc. of London* **6**, 84.

VAN DEN BERGHE, L. (1965) 'Le *Lantana camara* L. Nouveau fléau végétal en Afrique Orientale', *Bull. Acad. R. Sci. d'outre Mer*, Bruxelles 1123–9.

WALKER, A. R. P. (1953) 'Influence of parasitism on aspects of health' in *Internat. Rev. trop. Med.*, vol. 2, New York & London: Academic Press.

WALTON, G. A. (1962) *Symposium Zool. Soc. London* **6**, 83.

WANSON, M., COURTOIS, L. and BERVOERTS, W. (1950) 'L'extinction des simules de rivières à Léopoldville', *Ann. Soc. belge Med. trop.* **30**, 629–40.

WEBBE, G. (1969) 'Progress in the control of schistosomiasis', *Trans. R. Soc. trop. Med. Hyg.* **59**, 489–98.

WORLD HEALTH ORGANIZATION (1968) *Resistance of Malaria Parasites to Drugs*, Tech. Rept. Series 375, Geneva.

WILCOCKS, C. (1962) *Aspects of Medical Investigation in Africa* (Heath Clark Lectures, 1960), London: Oxford University Press.

WILLETT, K. C. (1960) 'Research on human trypanosomiasis 3. Animal reservoirs of *T. gambiense*', *Ann. Rept. West Afr. Inst. Tryp. Res.* 12–13.

WILSON, D. B. (1959) Report on the Pare-Taveta malaria scheme' Dar-es-Salaam: Government Printers.

WILSON, D. B. and WILSON, M. E. (1962) 'Rural hyperendemic malaria in Tanganyika Territory. II', *Trans. R. Soc. trop. Med. Hyg.* **56**, 287–93.

WOODRUFF, A. W., CHOJA, D. P., MUCI-MENDOSA, F., HILLS, M. and PETTIT, L. E. (1966) 'Onchocerciasis in Guatemala. A clinical and parasitological study with comparisons between the disease there and in E. Africa', *Trans. R. Soc. trop. Med. Hyg.* **60**, 707–34.

6

THE PARASITIC LIFE:
I PROBLEMS OF THE PARASITE

IN THE first chapter, the scope of parasitology was delimited without defining the parasite. This was intentional because the succeeding chapters dealt with the broad subject, considering the parasite in relation to its environment and not as a separate entity. Now it is time to discuss the parasite itself, and as a corollary it is desirable to consider also the observer. It is just as necessary to examine the relation of the parasite to the parasitologist as it is to the environment. In Chapters 6, 7 and 8 we shall deal with some of the current problems, and in the last chapter, with those of the past where the pattern became clarified in the course of history.

Parasites may be classified into various categories, including commensals, symbionts and true parasites, i.e. germs which injure the host, however small and undetectable may be the lesions. The dictionary definition of the word 'parasite' meets our broad needs: an animal or plant which lives in or upon another organism. We shall see how parasites evolved from free-living animals or plants, gradually became associated with a 'host' and penetrated more and more deeply into its tissues. The course of the parasite—and of the parasitologist—may be smooth, or it may be beset by problems some of which are overcome and others which prove insuperable. We shall examine examples of each.

Rogers and Somerville (1963) discuss how an organism passes from one environment to another; it is a hazardous process only to be crossed by a bridge and under the protection of various devices to enable the organism to withstand the adverse factors in the new conditions. A major problem is to cope with the diverse enzyme systems which are encountered, and perhaps form the ultimate obstacle to parasitism. The Australian

helminthologists point out that the invasive or protective stage
of a parasite is non-feeding and non-growing (note for instance
these features in the sporozoites of coccidia and of nematode
larvae) but on entry into the new host, neuro-secreted hormones
reactivate the larva, or other biochemical factors initiate the
development of the sporozoite.

Many parasites keep to as regular a time-table as do com-
muters on suburban railways. They are compelled to do so in
order to catch the bus to their next destination. Malaria
parasites and microfilariae provide excellent examples (see pp.
97 and 98). From the earliest times the strict periodicity of the
human species of *Plasmodium* was recognized in quotidian,
tertian and quartan malaria. Aristeus,[1] physician to Cotta, the
prefect of Alexandria, greatly admired the profound harmony
displayed by the quartan fevers; incidentally, he obtained
more enjoyment in studying such features of infectious diseases
than in curing them, an attitude still prevalent in some quarters
today! The synchronization of fertilization of parasites with the
rhythm of the host has been shown to exist also in certain
coccidians and their invertebrate hosts (polychaete worms);
Porchet-Henneré (1967) suggested that the hormones in the
adult stage of the latter stimulated the latent protozoan into
sexual activity just at the moment when the parasites are
discharged from the worm.

THE PARASITE'S PROGRESS

The free-living amoebae or amoebo-flagellates, *Hartmanella* and
Naegleria are a convenient starting point; these organisms are
found in soil or water and in recent years have been shown to
possess the property of parasitizing and killing man. Cases have
been reported from the U.S.A., Australia and Czechoslovakia
(Černa, 1966) and probably, under the name of *Iodamoeba*, from
New Guinea. Infection is thought to take place by violent
contact with infected water, as in diving into lakes or water-

---

[1] *A force d'étudier ce qu'on nomme les maladies, j'en suis arrivé à les considérer comme les
formes nécessaires de la vie. Je prends plus de plaisir à les étudier qu'à les combattre. Il y en a
qu'on ne peut observer sans admiration et qui cachent, sous une désordre apparent, des harmonies
profondes, et c'est certes une belle chose qu'une fièvre quarte!* From *Thaïs* by Anatole
France.

skiing on them; the organism is driven through the nose and cribriform plate into the brain where it produces meningo-encephalitis. The occurrence of the first case of this disease in England (Bristol) in a child only 2 years old suggests that water sports are not the sole medium of transmission (Apley *et al.*, 1970).

Culbertson *et al.* (1968) suggest, on the basis of results obtained by inoculating mice with cultures of *Hartmanella* and *Naegleria*, that a pulmonary infection may first arise which metastasises in the brain; the smaller species—*Naegleria*—seems to be more concerned in the fatal human cases. He also showed that several strains grow best at 37°C instead of the low temperatures at which free-living forms live; in fact Mrs Adam (1964) showed that 50 per cent of her strains of *Hartmanella castellanii* isolated from fresh water in Scotland also grew well at higher temperature.

Hartmanellosis may thus be regarded as an accidental parasitic infection of man and presumably demonstrates a method by which the gulf between free-living and parasitic organisms is traversed. Jirovec (1967) has described many examples in which this process can be experimentally induced, particularly of free-living ciliates into aquatic insects and also, after injury, into tadpoles. At one time, I thought that the free-living organism, *Euglena*, had parasitized the skin of people bathing in Lake Victoria where this organism was breeding extensively; they developed an itchy eruption in which I could find no trace of *Euglena* however, and when these protozoa were applied experimentally to the skin no rash appeared. This was forty years ago, and before schistosomiasis was known to occur in the locality—and probably the cercariae of *S. mansoni* were really responsible for the condition.

Entamoeba moshkowskyi is usually regarded as a free-living organism, and Meirovitch (1966) has speculated that some such amoeba was the ancestor of forms which ultimately invaded turtles and then reptiles (e.g. *E. invadens*). It is possible that *E. histolytica* also originated in this way, eventually reaching monkeys and then the higher apes. Urban troops of monkeys in India and Pakistan may thus constitute a zoonosis (Garnham, 1962).

Certain well-known parasites of man and animals have a

free-living existence which lasts much longer than their parasitic life in the vertebrate. *Histoplasma* persists in the dust of bat-haunted caves, *Leptospira* in the water of canals and *Clostridium tetani* in earth, while *Strongyloides* has a prolonged existence outside the body, and the cysts of many protozoa and helminths require exposure to the external environment to ensure their maturation. Only the larval stages of mermithid nematodes are parasitic (in the body cavity of various insects); the adults are free-living in water or earth.

Let us now turn to the ectoparasites, which are largely insects, ticks and mites. The most superficial contact occurs between mosquitoes and their vertebrate hosts, some being quite versatile in their diet—like *Mansonia uniformis* and feed on almost any vertebrate, while others are host specific, like *Anopheles dureni*, biting solely arboreal rodents.

The next in line is the tick. The soft ticks, e.g. *Ornithodorus moubata* live in cracks in the floor and walls of human habitations close to man, and bite him at frequent intervals. The hard ticks come closer in that they spend days attached to the skin of their hosts while feeding.

Sheep keds (hippoboscid flies) remain for most of their lives *in* the wool of the animal, while nycteribiids and strebliids of bats live permanently in their fur and take blood at all times. Lice approach even closer in that they stay on the skin during their adult life and live on the hair as nits.

Now parasitism advances to a further stage, because the parasite takes up residence *within*, instead of *on*, the skin. The scabies mite, *Sarcoptes scabei*, invades the skin all over the body; the jigger, *Tunga penetrans*, attacks particularly the feet and both these parasites become mature in the tissues of the host. Various flies grow as maggots in the skin or in open wounds, and produce the condition known as myiasis occasionally in man, but commonly in domestic animals and even more in the wild fauna. In Central America, *Dermatobia hominis* spends its larval life in the skin of wild animals such as the jaguar; the fly emerges and deposits eggs on the thorax of mosquitoes (e.g. *Janthinosoma* spp.); the eggs hatch and when the mosquito next bites, the larvae penetrate the skin at this point and grow to maturity. The original host lives in the forest, but the mosquitoes may carry the infection to cattle, which become riddled with the parasite and eventually to man, as a classical zoonosis.

These are localized conditions, but next, the parasite multiplies at the site of its introduction and diffuses through the skin, either in the immediate vicinity like *Leishmania* in an oriental sore, or reaching the mucous membranes in mucocutaneous leishmaniasis. *Onchocerca volvulus* goes deeper by invading the subcutaneous tissue in various parts of the body, and the microfilaria may even reach the eye. The leprosy bacillus quickly passes from the skin to the nasal mucosa, the peripheral nerves, the eye and sometimes reaches the internal organs (Edington and Gilles, 1969).

The truly endogenous life of the parasite begins when it enters the orifices of the body. It may then remain in unhealthy gums in the mouth like *Entamoeba gingivalis*, or in the genital tracts of man, like *Trichomonas vaginalis*; or the venereal parasites may penetrate more deeply like *Trichomonas foetus* into the uterus of the cow, or *Trypanosoma equiperdum* to the blood of the horse.

The portal of entry for the intestinal protozoa is the mouth and from there, the cyst is passed through to its optimal site of development somewhere in the alimentary tract. Biochemical and biophysical peculiarities of different portions of the gut in both vertebrates and invertebrates probably determine the actual site of invasion. The parasite may then multiply in the lumen, e.g. *Giardia intestinalis* in the duodenum or *Entamoeba coli* in the colon, or it may attack the mucosa and multiply in the submucosa; *Entamoeba histolytica* follows the latter course, but parasitization of the host may extend to all parts of the body if the organism enters the portal circulation. In most mammalian species of *Eimeria*, the infection is limited to the gut, but *E. stiedae* goes deeper, its sporozoites penetrate into the portal vessels and reach the bile duct epithelium in which they develop. In the fish species, the exact route that the sporozoite takes after leaving the gut is unknown, but final development of the parasites occurs in the internal organs (*E. clupearum* in the liver and *E. sardinae* in the testes).

We are now reaching the final stages of parasitism, when all or most of the internal organs become invaded. The parasite either enters the body through the unbroken skin like the cercariae of schistosomes and pursues a relatively simple route to the preferred site for their further development, or is conveyed more elegantly by the bite or ingestion of an invertebrate, and thence undergoes one of those incredible cycles of development, such as that of the malaria parasite or *Trypanosoma cruzi* in man, or *Aggregata eberthi* in the octopus.

Perhaps the brain may be regarded as the last stronghold of defence against the parasite, and this is eventually reached by a number of species, which inevitably end up in that organ. *Toxoplasma gondii* finds its final refuge there, as do *Trypanosoma gambiense* and *Borrelia duttoni*. No allusion has been made to the path of parasites in invertebrates, but it might be mentioned that the brain of the *Culicoides* midge is the terminal site of development of the common malaria parasite of monkeys, *Hepatocystis kochi*.

The ultimate degree of parasitism is reached when one parasite invades another, a condition known as superparasitism. Protozoal parasites are found in helminths (e.g. *Histomonas meleagridis* in *Heterakis gallinae*; microsporidians in *Toxocara* and *Moniezia*) as well as in other protozoa. The microsporidian *Brumptina* occurs in *Opalina* and bacterial killer particles in *Paramecium*. Organisms living on and in protozoa are described in great detail by Gordon Ball (1968) with the aid of a precise terminology. All parasites of vertebrates in their arthropod hosts (e.g. *Plasmodium vivax* in *Anopheles maculipennis*) may be regarded as examples of superparasitism.

This brief account of the first stage of parasite–host relationships has largely been limited to examples to which frequent reference has been made in this book, but enough have been discussed to show the progress from the free-living organisms to creatures which are so parasitic that they are parasites of parasites. We may now examine some of the difficulties that they meet on entry into the host. I have chosen the unusual happenings, as I believe that it is the *extraordinary* which is likely to have the most significance, just as the *unexpected* is often the best clue in research.

PROBLEMS OF THE PARASITE

The life cycle of many parasites involves obligatory residence in two or more types of host, usually one is an invertebrate and the other a vertebrate. The parasite will then have to face problems in both; it will have twice the difficulties to overcome as compared with a single-host parasite. A mosquito, living at a high temperature and infected with malaria parasites may not be able to survive long enough for the cycle of the *Plasmodium* to be completed and the parasite dies too. Or, the mosquito may have a double infection of malaria and a microsporidian, e.g. *Plistophora culicis* which inhibits the course of development of the *Plasmodium*. Or the mosquito itself may prove to be genetically unsuitable for the malaria parasite. In spite of such impediments, the transmission of a parasite through a vector has two great advantages: (1) it permits multiplication of the organism which is then diffused over the wide area traversed by the host and (2) in parasites which undergo a sexual cycle, favourable genetic recombinations are made possible.

The problem of how to get from one host to another always presents difficulties for the parasite (see Reed, 1970), and if a suitable vector is absent, transmission will cease and the infection will die out. This is perhaps the reason why *Plasmodium ovale* has failed to become established anywhere in the world other than in its natural home in tropical Africa; the infection has appeared temporarily in a few other places such as New Guinea and the Middle East; but, perhaps because there was not the right species of *Anopheles* present, *P. ovale* quickly disappeared. There is another factor however which favours the wide dispersal of *P. ovale*; this parasite often shows a delayed incubation period, during which time the patient may have travelled thousands of miles and the place where the diagnosis is made is very different from the region where the infection was contracted.

If a biting insect is involved in transmission, it will only become infected when infective stages of the parasite are present in the peripheral blood or tissues of the host. Hawking (1967 and 1968) has suggested the devices which the parasite employs to meet this need. Mosquitoes have a well-marked circadian rhythm in their biting activities; the parasites have similar rhythms in their prevalence in the blood. The peak of both should coincide in any host-parasite combination in order to ensure maximum infection. Thus, the microfilariae of *Wuchereria bancrofti* are retained in the pulmonary capillaries during the day, but at night when the mosquitoes bite, the raised oxygen tension allows them to escape into the peripheral circulation and the beneficial nocturnal periodicity is established. The microfilariae of *Loa loa* provide an interesting example in that the vector of the human strain, *Chrysops silacea*, is a day-feeder corresponding to the diurnal periodicity of the microfilariae, while the vector of the simian strain, *C. centurionis* feeds at night when the microfilariae accumulate in the peripheral blood of the monkey (Duke, 1959).

The gametocytes of a few species of *Plasmodium* are said to be produced rhythmically as part of the periodic (24, 48 or 72 hours') cycle of schizogony and become mature only for a few hours in the middle of the night at the time when the vector is most active. In these species, which may be the exception rather than the rule, Hawking suggests that the parasites take their 'cue' from the temperature cycle of the host. The actual stimulus may be the varied secretion of adrenocorticosteroids in the course of the 24 hours (Bliss *et al.*, 1953) or the periodic availability of DNA or its precursors in the circulation which are required for the final nuclear divisions of the parasite.

Malnutrition

In order to live, the parasite has to feed, and the principal reason for its adopting the parasitic form of life is that nutriment becomes immediately available from the body of the host. If the host is starved, the parasite suffers from malnutrition, and in conditions of famine of a human population, the virulence of many infectious diseases diminishes. Malaria epidemics lessen, and poorly nourished people or those in the final stages of cancer cannot easily be infected with *Leishmania donovani*.

Perhaps the most striking example of the effect of malnutrition of the animal on the multiplication of the parasite is provided by restricting the host's diet to milk. Infections of *Plasmodium berghei* which are normally fatal to mice, or of *P. knowlesi* to rhesus monkeys, can be inhibited by placing the animal on a milk diet. Milk contains practically no para-aminobenzoic acid, and the erythrocytic forms of the malaria parasite are faced with the problem of folic acid synthesis which requires this substance and without which they cannot grow. The infection becomes stationary or may even be entirely suppressed. Exoerythrocytic stages of *Plasmodium*, on the other hand, do not need this substance directly and tissue schizonts of *P. cynomolgi* grow normally in monkeys on such restricted diets (Garnham, 1966). Although *Toxoplasma gondii* is a tissue parasite, unlike the malaria parasite it is unable to get its para-aminobenzoic acid from cells in which it lives, and the course of toxoplasmosis in mice on a milk diet is inhibited.

Some parasites are more affected than others by different nutritional defects in the host, and Kretschmar (1968) attempts to summarize the experimental work on (a) intestinal, (b) blood, and (c) intracellular protozoal parasites, to illustrate the variety of changes which may occur; the parasite may be injured by interfering with its metabolism without damaging the host.

Competition for food

Polyparasitism is extremely common in nature, and both vertebrate and invertebrate animals may be infected simultaneously with different viruses, bacteria, protozoons and helminths. These may have to compete with each other for food, and if unsuccessful, their growth may be entirely suppressed or much limited.

There are many examples of such conflicts in the blood of the host. The spirochaetes often have an adverse influence on blood protozoa. Thomson and de Muro (1932) showed that *Borrelia duttoni* depresses an infection of *Trypanosoma rhodesiense* in mice and Galliard *et al.* (1958) observed a similar effect with other species of Borrelia (*B. merionesi*, *B. microti* and *B. turicatae*). Tate (1951) demonstrated the antagonism between *Spirillum minus* and both *Trypanosoma lewisi* and *T. equiperdum*, and van Thiel (1964) between this *Spirillum* and *T. cruzi*. *Eperythrozoon coccoides* hinders the development of *P. berghei* in mice (Peters, 1965), a complex relationship which has been shown by Ott, Astin, and Stauber (1967) to be due to several causes, including the competition for metabolites, stimulation of the reticulo-endothelial system and production of reticulocytes.

Experiments with double infections of *Babesia* spp. and rodent malaria parasites gave some unexpected results in regard to what was interpreted as cross immunity by Herbert Cox (1968) and by Frank Cox (1968). Mice which had recovered from infections of *Plasmodium chabaudi* or *P. vinckei* were shown to be nearly insusceptible, a fortnight later, to *Babesia rodhaini* or *B. microti* and vice versa. The so-called cross-immunity may have been due to the persisting effect of a competition for nutrients, or more probably to the continued presence of 'interferon' which barred the invasion of the cells by the second parasite (see below). The latter is a non-specific effect and seems more likely to be the explanation than a classical cross-immunity between two totally unrelated organisms. There are various examples of viruses having an adverse effect on malaria parasites (ornithosis virus on *P. lophurae*—Jacobs, 1957; Semliki Forest virus on *P. gallinaceum*—Bertram *et al.*, 1967; etc.). The foregoing double infections are all of an experimental origin and one wonders how often the depressant effect occurs under natural conditions. Wijers (1969) recently attempted to explain the rarity of *T. rhodesiense* infections in cattle (see p. 82) by the fact that these animals are often infected at the same time by *T. vivax*, a much more virulent parasite which dominates the blood picture and inhibits the more slowly growing polymorphic trypanosome. He also suggested that reptilian blood (on which *G. fuscipes* chiefly feeds) may have an adverse effect on the development of the trypanosome in the gut of the fly.

An ingenious explanation for the inhibitory effect of a virus on the development of malaria parasites has recently been proposed by Jahiel *et al.* (1968) who found that mice which had been inoculated with sporozoites of *P. berghei* and 24 hours later with Newcastle Disease Virus easily overcame the malaria infection—probably because of the heavy production of interferon induced by the virus. Similarly Remington and Merigan (1968) demonstrated the inhibitory effect of interferon on the growth of *Toxoplasma gondii* in tissue culture.

It must not be thought that the presence of a second parasite necessarily interferes with the development of the first, and there are many examples of the exacerbation of one infection by another. The virus of distemper activates toxoplasmosis in dogs and renders them more susceptible to the disease; a hookworm infection has a similar effect. A double infection in mice of Rauscher leukaemogenic virus and *Plasmodium berghei yoelii* results in a 100 per cent mortality from the malaria (Salaman *et al.*, 1969). A tuberculous peritonitis can light up a latent infection with *Balantidium coli* and cause a fatal balantidial dysentery. In such cases, the resistance of the host is lowered, by blockade or 'paralysis' of the defence mechanisms, and the parasite is able to multiply unhindered.

If a man is infected simultaneously with the sporozoites of *Plasmodium vivax* and *P. falciparum*, the *P. falciparum* develops normally in the liver and gives rise after an incubation period of about ten days to a patent infection; *P. vivax* remains dormant and only appears in the blood stream after some months. The explanation of the occultation of the benign by the malignant tertian parasite remains absolutely obscure, but has been discussed in some detail by Shute (1946) who made these observations. He noted also that if the *vivax* sporozoites were given a 3 or 4 days' start over the *falciparum*, both infections appeared simultaneously in the blood.

Protection from host's defences

The fundamental problem of the parasite is to overcome the resistance of the host. In the words of Sergent (1963), there may be three results of invasion: (1) the parasite multiplies without restraint and kills the host, and itself becomes extinguished, (2) the host destroys the parasite, or (3) multiplication of the parasite occurs on a reduced scale, and host and parasite eventually settle down together in the state of premunition. The last result is the ideal one from the point of view of the parasite and possibly also of the host. It may well be that even from the human angle, the establishment of premunition is the best answer to the problem of many infectious diseases, and that the emphasis, today, on eradication is both unnecessary and unwise: the germ-free animal is not the most healthy!

The simplest example of how an organism overcomes a hostile environment is the process of encystment; the thick wall of the cyst suffices to protect the parasite against desiccation. When conditions change, e.g. when the cyst returns to the alimentary tract, the wall is dissolved and the parasite resumes its active existence. Similarly, spores with resistant walls are formed by certain bacteria and protozoa which enable them to survive, perhaps for years, under adverse conditions.

The parasite has to contend with natural immunity. Apart from the general unsuitability of certain host-parasite relations, in some instances, a host which is normally excellent—like man for rodent species of malaria parasites—at other time presents difficulties for the parasite because of genetic abnormalities. Thus, multiplication of *Plasmodium falciparum* is hindered in people who possess sickle cell haemoglobin (Allison, 1954) or have a genetic deficiency of the enzyme, glucose 6 phosphate-dehydrogenase (Gilles *et al.*, 1967). Haemoglobin S differs from normal haemoglobin in possessing valine instead of glutamic acid in the B chain of the molecule; this is said to increase the viscosity of the cytoplasm of the erythrocyte and interfere with its uptake by the parasite. This ingenious explanation is invalidated because Raper (1959) showed that the parasite had no difficulty in growing in cultures containing sickle cells. The probable cause of this curious phenomenon is that the corpuscles containing *P. falciparum* retreat to the internal organs where the oxygen tension is lower than in the peripheral blood, sickling follows *in vivo*, and the schizont is expelled before it reaches maturity.

Onyango (1968) has shown that Africans of blood groups other than B are relatively insusceptible to *Trypanosoma rhodesiense*.

A most inexplicable condition is the almost total immunity of the West African to *Plasmodium vivax*. Why should this parasite and no others, except for the closely related *P. cynomolgi* and *P. schwetzi*, be unable to develop in the blood of the true Negro, and yet *P. vivax* grows perfectly well in any other African (Bantu) and any other race of mankind? Thus, in West Africa, but nowhere else, *P. vivax* is missing from the local pattern of species of malaria parasites. The only exception is to be found in countries colonized by the West African Negro, viz. Surinam in South America and Haiti in the Caribbean where *P. vivax* is similarly absent, and in the U.S.A. where the Negro inhabitants have been shown to be largely insusceptible to this parasite. If, however, there is much admixture of Caucasian blood, the parasite manages to survive, and the person of mixed race is no longer resistant to this form of malaria. The reaction of the Negro to this group of parasites provides an interesting guide to their phylogenetic relationships and indicates the affinity of *P. vivax*, *P. cynomolgi* and *P. schwetzi*, but the lack of relationship with *P. ovale*, to which the Negro responds normally.

The basis of the establishment of a parasite in a given host is host-parasite specificity; if the relationship is perfect, as in the last category of Sergent's triad (see above p. 101), the parasite has no problem; if natural immunity or innate resistance is complete, then the problem is insuperable. but under certain conditions this obstacle can be overcome. These conditions are largely experimental and have been analysed by Zuckerman (1968) in a recent review on specificity of parasites in the verte-brate blood.

The spleen is the all-powerful organ which protects the host from the danger of multiplication of the parasite. In many potential parasitic infections, the splenic barrier proves too great a problem for the organism to surmount; but occasionally the parasite is in luck if by chance the spleen is non-functioning or actually absent. The immunogenic cells may have been blocked, in nature perhaps by a virus, or experimentally by repeated inoculations of carbon particles; then, if the animal is inoculated with a parasite (e.g. rats with *Plasmodium berghei yoelii*) which is normally non-lethal, the parasite is able to multiply without hindrance and kills the host.

An even more striking effect is produced when a normally harmless parasite gets into a host *without* a spleen. There have been three dramatic occurrences in the last few years, where men who had lost their spleens by operation became exposed to piroplasms—protozoal parasites of cattle and other animals which normally do not affect man. Transmission takes place by tick bite, and in these three cases, very severe infections occurred, two of which were fatal. The cases were originally diagnosed as malaria and the parasites were only later correctly identified. The first was from Zagreb in Jugoslavia (Skrbalo and Deanovič, 1957) and was due to *Babesia bovis*; the man was a farmer who eleven years previously had had a car accident and his spleen had to be removed. He was admitted to hospital with fever, jaundice and haemoglobinuria; 'rings' were found in his blood and these were identified as *Plasmodium falciparum*. The patients died eight days later and the case was diagnosed as blackwater fever. This disease had not been seen in Jugoslavia for many years and the films were therefore sent to various protozoologists (Geigy, Swellengrebel and the writer), all of whom stated that the organisms were a species of *Babesia*. An investigation of the farm showed that the fields were heavily infested with ticks and the cattle had red water fever.

The second case occurred in California and has been described by Scholtens *et al.* (1968). The man was a keen photographer and spent much of his spare time in isolated coastal areas near San Francisco. Early in June 1966, he became severely ill with fever and chills and was admitted to hospital. Parasites resembling malaria 'rings' were seen in the blood films; malaria was diagnosed and he was treated with chloroquine (250 mg weekly for 15 weeks). The history revealed no recent exposure to malaria and the blood films were sent to the National Communicable Diseases Center at Atlanta for confirmation. The parasites were found to contain no pigment and were sometimes in the Maltese Cross form, typical of certain piroplasms. It was then discovered that a splenectomy had been performed on the man two years previously, on account of congenital sphaerocytosis. Serological tests were positive for piroplasmosis but negative for malaria. It seems probable that the infection had originated in wild rodents or possibly deer, and was due to some species of *Nuttallia*.

The third case was contracted in Ireland (Fitzpatrick *et al.*, 1968) and was due to *B. divergens*. The details are given on pp. 112ff. below.

With a spleen, man seemed to be immune to these parasites; without it, he is as susceptible as if the disease were plague. The occurrence of another case in the United States (Benson *et al.*, 1969) in a woman *with* a spleen invalidates this conclusion, unless it is assumed that the function of the organ in this instance was defective. The patient had been bitten by ticks, probably from rodents, in Nantucket Island and 6 weeks later developed a febrile illness, accompanied by 'rings' in the blood, which on inoculation into hamsters, gave rise to a typical babesial infection. Like the Californian case, she recovered, and it seems as if the rodent piroplasms may be less virulent to man than the species from cattle. It is strange that comparatively few natural infections of piroplasms have been reported in the New World.

Throughout parasitology, the spleen represents a thorn in the flesh of most parasites! It plays an equal role in overcoming infections accompanied by acquired immunity, though now other tissues are also involved.

Acquired immunity operates by the sensitization, hypertrophy and secretions (antibodies) of the lymphoid-macrophage system. If the parasites multiply very quickly this mechanism starts too late and the host dies. But if the rate of increase is slower, immunity begins to take its toll of the parasite, which may react to the inimical process by antigenic variation; at least this is the mode of defence employed by certain organisms in the blood, such as relapsing fever spirochaetes, trypanosomes and malaria parasites. The new variants can multiply until the host produces suitable antibodies; a few mutants survive however and fresh variants are produced at intervals. New work on so-called 'relapse variants' has been carried out by Brown and Brown (1966). They showed that in drug-treated *P. knowlesi* malaria in monkeys, a variant managed to survive which gave rise at a later date to a 'relapse' (better termed a 'recrudescence'); the process was indefinitely repeated, each variant stimulating the production of new specific agglutinins, which destroyed all the current brood of parasites.

Another protective device of the parasite is to take refuge in some site in the body which is unexposed to the effects of immunity. The secondary exoerythrocytic cycles of malaria parasites in the parenchyma cells of the liver are uninfluenced by the immunity which is built up against the blood stages, and the healthy appearance of such schizonts in the liver is quite remarkable, considering how their progeny—the merozoites— are immediately destroyed when they enter the blood stream (Garnham and Bray, 1956). The liver cycle continues until the time comes when the host loses its immunity and the parasites again are able to multiply in the red blood cells. Another type of malaria parasite, *Leucocytozoon simondi*, is present in the blood of wild duck, chiefly during the summer months when the vector fly, *Simulium venustum*, is active; then, during the winter, the parasite hibernates in an exoerythrocytic cystic form in the heart or lungs of the bird, only to emerge again in the spring (Desser *et al.*, 1968).

These latent (exoerythrocytic) forms of malaria parasites stimulate no adverse reaction by the host and usually the organism has no effect on the latter's tissues except to enlarge the cell in which it is living. Some species have however a remarkable influence on the nucleus of the host cell (see Garnham, 1966). *P. malariae* causes the nucleus of the parenchyma cell of the liver in which it is growing to double in size; *Hepatocystis kochi* and *H. fieldi* stimulate both hypertrophy and division of the nucleus; *Leucocytozoon* spp., has the most extraordinary effect, the nucleus of the host cell becoming as much as a hundred times the normal volume. It is interesting to watch the growth of the nucleus of the leucocyte or erythrocyte after the cell has become invaded in the blood stream; the tiny merozoite of *L. simondi* almost immediately provokes enlargement of the nucleus, a feature which is easier to detect than the parasite itself, and growth steadily progresses. The nature of the factor responsible for the gigantism is quite unknown and presents an interesting problem for study by the biochemical parasitologist. The phenomenon is not confined to haemosporidians, but is seen also in coccidian and microsporidian infections, in *Aggregata*, etc.

Gigantism has been explained as an expedient of the parasite for providing a source of food, the hypertrophied nucleus of the host cell supplying all the metabolites necessary for growth. It represents quite a sophisticated way of overcoming a fundamental problem.

Antiparasitic drugs

Parasites of man and his domestic animals have often to contend today with drugs used in the treatment of the resultant maladies. Many synthetic preparations exert their effect by blocking the metabolic pathways in the nucleus and cytoplasm of the parasite. Of essential importance in glucose metabolism is the enzyme chain: para aminobenzoic acid—folic acid—folinic acid. A drug like pyrimethamine becomes substituted in this linkage, the mechanism of biosynthesis is interrupted, and the parasite will die, unless it can by-pass the blockage. It may manage to do this by a slight shift in its metabolism, growth continues and the parasite is now pyrimethamine-resistant. Other drugs, such as quinine or the 4 amino-quinolines (e.g. chloroquine), act on multiple links, or affect the lysosomes in the cytoplasm of the parasite and the parasite finds it more of a problem to provide new routes for the varied biosynthetic processes involved.

The genetics of this phenomenon have been much studied in drug-resistant bacteria, and are said to be due to cytoplasmic inheritance of the new gene. Yoeli (1968) has recently put forward an interesting modification of this theory in regard to protozoa. A single erythrocyte may be invaded by two malaria parasites, one of which is drug resistant and the other is susceptible; they lie close together in the cell and feed by pinocytosis. This process involves the grabbing of ultramicroscopic portions of the adjacent tissue which in this instance is the neighbouring parasite. The susceptible parasite grabs a piece of the resistant one with which it has been in contact. Part of the cytoplasm of the resistant thus becomes incorporated in the body of the susceptible parasite, and the gene responsible for resistance is passed on to the progeny of this organism, in the so-called process of synpholia. So far, this theory remains unconfirmed.

REFERENCES

ADAM, K. (1964) 'The amino-acid requirements of *Acanthamoeba* sp. Webb', *J. Protozool.* **11**, 98–100.

ALLISON, A. C. (1954) 'Protection afforded by sickle-cell trait against subtertian malaria infection', *Br. med. J.* **1**, 290–4.

APLEY, J. et. al. (1970) 'Primary amoebic meningoencephalitis in Britain', *Br. med. J.* **1**, 596–9.

BALL, G. H. (1968) 'Organisms living on and in protozoa' in *Research in Protozoology*, vol. 3, pp. 566–718, Oxford: Pergamon Press.

BENSON, G. D., GALDI, V. A., ALTMAN, R. and FIUMARA, N. J. (1969) 'Human babesiosis—Massachusetts', *Morbidity and Mortality* **18**, 277–8.

BERTRAM, D. C., VARMA, J. G. R. and BAKER, J. R. (1964) 'Partial suppression of malaria parasites and of the transmission of malaria in

Aedes aegypti L, doubly infected with Semliki Forest virus and *Plasmodium gallinaceum*', *Bull. Wld Hlth Org.* **31**, 679–97.

BLISS, E. L., SANDBERG, A. A., WILSON, D. H. and EIK-NESS, K. (1953) 'The normal levels of 17-hydroxycorticosteroids in the peripheral blood of man', *J. clin. Invest.* **32**, 818–23.

BROWN, K. N. and BROWN, I. N. (1966) 'Antigenic variation in simian malaria', *Trans. R. Soc. trop. Med. Hyg.* **60**, 358–63.

CERVA, Z. (1966) 'Use of fluorescent antibody technique to identify pathogenic hartmanellae in tissue of experimental animals', *Folia Parasitol.* **13**, 328–31.

COX, F. E. G. (1968) 'Immunity to malaria after recovery from piroplasmosis in mice', *Nature* **219**, 646.

COX, H. W. and MILAR, R. (1968) 'Cross-protection immunization by *Plasmodium* and *Babesia* infections of rats and mice', *Am. J. trop. Med. Hyg.* **17**, 173–9.

CULBERTSON, C. G., ENSMINGER, P. W. and OVERTON, W. M. (1968). 'Pathogenic *Naegleria* sp.—study of a strain isolated from human cerebrospinal fluid' *J. Protozool.* **15**, 353–63.

DESSER, S., FALLIS, M. and GARNHAM, P. C. C. (1968) 'Relapses in ducks chronically infected with *Leucocytozoon simondi* and *Parahaemoproteus nettionis*', *Canad. J. Zool.* **46**, 281–5.

DUKE, B. O. L. (1959) 'Studies on biting habits of *Chrysops*. VI', *Ann. trop. Med. Parasit.* **53**, 203–14.

EDINGTON, G. M. and GILLES, H. M. (1969) *Pathology in the Tropics*, London: Edward Arnold.

FITZPATRICK, J. E. P., KENNEDY, C. C., McGEOWN, M. G., OREOPOULOS, D. G., ROBERTSON, J. H. and SOYANNWO, M. A. O. (1968) 'Human case of piroplasmosis (Babesiosis)', *Nature* **217**, 861–2.

FRANCE, A. (1844–1924) *Thais*, Paris: Calman Lévy.

GALLIARD, H., LAPIERRE, J. and ROUSSET, J. J. (1958) 'Specific behaviour of different species of *Borrelia* in the course of a mixed infection with *Trypanosoma brucei*. Its use as a test for identification of recurrent spirochaetes', *Ann. Parasitol. Humaine et Comp.* **33**, 177–208.

GARNHAM, P. C. C. (1962) 'Parasitological problems in tropical medicine' in *Radioisotopes in Tropical Medicine*, Vienna: IAEA.

GARNHAM, P. C. C. (1966) *Malaria Parasites and other Haemosporidia*, Oxford: Blackwell Scientific Publications.

GARNHAM, P. C. C. and BRAY, R. S. (1956) 'Influence on immunity upon the stages (including late exoerythrocytic schizonts) of mammalian malaria parasites', *Rev. Brasil. Malariol.* **8**, 151–8.

GILLES, H. M., WATSON-WILLIAMS, E. J. and TAYLOR, B. G. (1967) 'Glucose-6-phosphate dehydrogenase deficiency, sickling and malaria in African children in South Western Nigeria', *Lancet* **i**, 138.

HAWKING, F. (1967) 'The 24-hour periodicity of microfilariae: biological mechanisms responsible for its production and control', *Proc. R. Soc. B.* **169**, 59–76.

HAWKING, F., WORMS, J. and GAMAGE, K. (1968) '24- and 48-hour cycles of malaria parasites in the blood: their purpose, production and control', *Trans. R. Soc. trop. Med. Hyg.* **62**, 731–65.

JACOBS, H. (1957) 'Effect of ornithosis on experimental fowl malaria', *Proc. Soc. expl Biol. Med.* **95**, 372–3.

JAHIEL, R. I., VICFEK, J., NUSSENSWEIG, R. and VANDERBERG, J. (1968) 'Interferon inducers protect mice against *Plasmodium berghei* malaria', *Science* **161**, 802–4.

JIROVEČ, O. (1967) Le parasitisme artficiel des protozoaires libres', *Ann. Parasit. hum. Comp.* **42**, 133–40.

KRETSCHMAR, W. (1968) 'Der Einfluss der Wirternährung auf die Entwicklung von parasitischen Protozoen', *Z. Parasit.* **31**, 20–37.

MEIROVITCH, E. (1968) 'Pathogenicity of amoebae other than *Entamoeba histolytica*', *Proc. 8th Cong. trop. Med. Malaria*, Teheran.

ONYANGO, R. J. (1968) in *Symposium on African Trypanosomiasis*, London: Ministry of Overseas Development.

OTT, A. J., ASTIN, J. K. and STAUBER, L. A. (1967) '*Eperythrozoon coccoides* and rodent malaria: *Plasmodium chabaudi* and *Plasmodium berghei*', *Exper. Parasit.* **21**, 68–77.

PETERS, W. (1965) 'Competitive relationship between *Eperythrozoon coccoides* and *Plasmodium berghei*, in the mouse', *Exper. Parasit.* **16**, 158–66.

RAPER, A. B. (1959) 'Further observations on sickling and malaria', *Trans. R. Soc. trop. Med. Hyg.* **53**, 110–17.

REED, C. P. (1970) *Parasitology and Symbiology*, New York: Ronald Press Co.

REMINGTON, J. S. and MERIGAN, T. C. (1958) 'Interferon: protection of cells infected with an intracellular protozoon (*Toxoplasma gondii*)', *Science* **161**, 804–6.

ROGERS, W. P. and SOMMERVILLE, R. I. (1963) 'The infective stage of nematode parasites' in *Advances in Parasitology*, vol. 1, Ed. Ben Dawes, London: Academic Press.

SALAMAN, M. H., WEDDERBURN, N. and BRUCE-CHWATT, L. (1970) 'The immunodepressive effect of a murine plasmodium and its interaction with murine oncogenic viruses', *Exper. Parasit.* In Press.

SCHOLTENS, R. C., BRAFF, E. H., HEALY, G. R. and GLEASON, N. (1968) 'A case of babesiosis in man in the United States', *Am. J. trop. Med. Hyg.* **17**, 810–13.

SERGENT, E. (1963) 'Latent infection and premunition' in *Immunity to Protozoa*, Eds. Garnham, Roitt and Pearce, Oxford: Blackwell Scientific Publications.

SHUTE, P. G. (1946) 'Latency and long-term relapses in benign tertian malaria', *Trans. R. Soc. trop. Med. Hyg.* **40**, 189–200.

SKRBALO, Z. and DEANOVIČ, Z. (1957) 'Piroplasmosis in man, Report of a case', *Doc. Med. Geogr. Trop.* **9**, 11–16.

SPRENT, J. F. A. (1963) *Parasitism*, London: Baillière, Tindal & Cox.

TATE, P. (1951) 'Antagonism of *Spirillum minus* to *Trypanosoma lewisi* and *Trypanosoma equiperdum*,' *Parasitology* **41**, 117–24.

VAN THIEL, P. H. (1964) 'L'antagonisme entre *T. cruzi* et *Sp. minus*,' *Ann. Soc. belge Méd. trop.* **44**, 347–51.

THOMPSON, J. G. and DE MURO, P. (1932) 'The influence of *Treponema duttoni* on an infection with *Trypanosoma rhodesiense* in mice', *J. trop. Med. Hyg.* **35**, 33–38.

WIJERS, D. J. B. (1969) 'The possible rôle of domestic animals in the epidemiology of *T. rhodesiense* sleeping sickness', *Bull. Soc. Path. exot.* **62**, 334.

YOELI, M., UPMANIS, R. S. and MOST, H. (1969) 'Drug resistance transfer among rodent plasmodia', *Parasitology*, **59**, 429–47.

ZUCKERMAN, A. (1968) 'Basis of Host Cell-Parasite Specificity' in *Infectious Blood Diseases of Man and Animals*, 1, 23–36, New York: Academic Press.

7

THE PARASITIC LIFE:
II. PROBLEMS OF THE
PARASITOLOGIST

In 1947 the late Professor R. Leiper asked the writer if he thought it were wise to train so many young parasitologists, because he felt that in ten or twenty years there would be no jobs for them. Not only has his prophecy proved incorrect, but he himself still worked (when these lectures were given) at the Commonwealth Bureau of Helminthology at the age of 83. By sheer numbers, we are still overwhelmed by the animal parasites: 500 million cases of ancylostomiasis, 100 million of schistosomiasis and 100 million cases with a million deaths per annum of malaria. In Ceylon alone where the disease was thought to be nearly eradicated, there were 100 000 cases of vivax malaria diagnosed in 1968 and probably a total of a million. These are just three of many common parasitic diseases of man; there is an equal number among the domestic animals, and while we are faced with a problem on this scale, surely there can scarcely be enough parasitologists. An indication of the advance of parasitology, *sensu strictu*, in Great Britain is the progress of the British Society of Parasitologists, which was formed in 1961, largely through the inspiration of Ann Bishop, F.R.S., its first President. It has now a membership of over 700, the majority of whom are young and present papers of much interest to the two meetings held each year, one in London and the other at different universities from one end of the kingdom to the other.

Two years ago, Browning (1967) took a look at progress in pathology and was delighted at the great increase in numbers of pathologists and the recognition of the subject as an independent scientific discipline. He raised the question of the desirability of a medical qualification for a pathologist and indicated that its absence would be a hindrance in dealing with patients.

A similar problem faces the parasitologist, who fortunately is not tied to clinical medicine, but has open before him the field of general epidemiology, veterinary science, and zoology as a pure discipline. Nevertheless, probably a medical training provides the best introduction to our subject; but it is only a preliminary, and it is essential, even today with all its specialization, to learn also parasitology, entomology and epidemiology. Later we can become experts in our own 'mini-fields'.

Scientific problems

The first problem that faces the parasitologist is to select suitable material and to decide what to do with it. He has to find the right locality for his work and an organization to back it. If he is medical, he must divert his gaze from Harley Street and focus on a more distant horizon. His early career should better be spent in the field, and the tropics still offer the best prospects, if only because there are more species of animals, including parasites, present in the warmer climates. In the temperate zone, the seasonal variations of the environment inhibit the proliferation of species and fewer are able to survive (Mayr, 1962).

Orthodox science regards the modern laboratory as the proper place of work and Hans Krebs (1967) warned research workers not to travel too much abroad. But the investigation of the zoonoses entails working in the field, and comparing conditions in one place with those in another.

Nevertheless, there are parasitological problems everywhere, even on our own doorstep. In the London School of Hygiene and Tropical Medicine, we have Woodruff working on that important, though aberrant parasite of man, *Toxocara*, and Lumsden on trypanosomes in Highland cattle; Bruce-Chwatt on imported parasites in the Pakistanis and Africans of Bradford; Ormerod on the ectoparasites of ponies in the New Forest; Baker on the malaria parasites of birds and Shortt on the piroplasms of small mammals at the St Albans field station; Varma on ticks in Scotland and Nelson on *Trichinella* in Ireland. A recent experience of mine (Garnham *et al.*, 1969) involved a trip to Ireland, to make a few observations on the case of human piroplasmosis that has been mentioned above (p. 104).

The visit illustrates some facets of the parasitologist's life and the investigation is briefly related below.

A man (F.) went on holiday in August 1967 to Lough Corrib in Galway (Plate IV, Fig. 14), and camped with his family on its western shore. The fields around the camp are alive with ticks dropping off the cattle; the cattle here suffer from red water fever (caused by the piroplasm *Babesia divergens*). F. spent his time resting in the fields, for he was still not fully recovered from a severe operation for a duodenal ulcer which he had undergone four months earlier. Owing to adhesions, the spleen had to be removed in the course of the operation. F. was unaware of his peril, for the splenectomy had rendered him vulnerable to the cattle infection, and he must have been constantly exposed to the bites of ticks during the five days he was there. On return home, he became severely ill with fever, jaundice and anaemia, and malaria-like rings were found in his blood. He had received a blood transfusion after the operation earlier in the year and transfusion malaria was suspected. He died the next day, before the parasite had been correctly identified. The blood film was sent to the Malaria Reference Centre at Horton Hospital, where I diagnosed the parasite as a piroplasm and not a malaria parasite. The Irish doctors (Fitzpatrick *et al.*, 1968) then investigated the case in detail.

I was interested in pursuing the matter further, for I had been concerned in the first case of human piroplasmosis from Zagreb. I had come to the conclusion that people with intact spleens and bitten by infected ticks should harbour, at some time or other, occult parasites in their circulation, as happens so frequently in other parasitic conditions (see Nicolle's work on this subject, p. 117). Such parasites should be most easily demonstrable by inoculation of heparinized blood into splenectomized calves.

The Irish case seemed to present an excellent opportunity to put the theory to the test and I went to the site exactly a year later, where I was met by Dr Cotton Kennedy, the pathologist who had investigated the case, and by Harry Hoogstraal and Gerald Walton, the tick experts. The regional medical officer (Dr MacCon) and the veterinary officer (Patrick Fagan) helped us with our work and took us to the place where F. had camped.

Just before reaching the camp, we saw a large gathering in a field: it was the Irish Army participating in the new MGM film on Alfred the Great. The men had been there for about a fortnight, spending much of their time sun-bathing on the grass and inevitably bitten by ticks. Our experiment was done for us. Instead of trying to find some Irish peasants working in these fields and to persuade them to let us take their blood, we seized the opportunity that lay before us and asked the commanding officer to provide volunteers for blood samples. He at once agreed, gave 10 ml of his own blood and produced 35 soldiers also. The samples were mixed in suitable containers in the travelling laboratory of Dr Kennedy, parked on the site, and were then transferred to a large thermos flask containing ice. The chilled blood was flown from Shannon to London Airport where Mr Donnelly of the Central Veterinary Laboratory, Weybridge, met the plane, and within 24 hours of taking, the material had been inoculated into two splenectomized calves. We also collected ticks, nymphs and larvae (*Ixodes ricinus*) from the vicinity and these were later taken to Weybridge by Dr Hoogstraal and fed on other splenectomized calves.

The end of this long story is a negative result; at least no babesial infections arose in the calves. However, one calf which had been bitten by the *Ixodes* developed 'bovine tick-borne fever', a rickettsial disease confined to the British Isles and Scandinavia (Tuomi, 1966). When this animal and the other calves were later challenged with *B. divergens* (to confirm that a subpatent infection of the piroplasm had not occurred) the others all developed severe red water fever, but the rickettsial calf exhibited only a mild infection. My wife, who also took part in this work by collecting and sorting the ticks, immediately suggested the possibility of using the fairly harmless rickettsia as a vaccine against the more virulent red water fever.

I still feel that subpatent infections of babesiosis *must* exist in agricultural workers throughout the world, although the Irish experiment, and other attempts that I have made, have failed to establish this theory. Dr Hoogstraal and other workers in NAMRU III in Cairo intend to investigate the subject intensively in the inhabitants of the Delta, where this disease is common in the local cattle, and where many people have been splenectomized for chronic bilharziasis.

There are various lessons to be learnt from this experience; amongst others, the importance of securing local co-operation and of team work of experts in different lines, and perhaps also of never abandoning a theory until you have tested it in all directions

The danger of splenectomy for farm workers and others should be properly realized (see Cooper, 1969). There is a specific occupational hazard in human piroplasmosis because in unsuitable conditions harvesters and other agricultural machinery are unstable, liable to overturn and crush the driver; the spleen is ruptured and the organ is removed on the operating table. The victim is now under severe risk, and he may contract piroplasmosis, a disease difficult to recognize; if it is diagnosed in time, probably the best drug to use would be Amicarbalide (= Diamprin, M and B) at a dose of 10 mg per kg, or perhaps prolonged treatment with chloroquine.

The subject of piroplasmosis in primates can fortunately be studied in animal models, and directly I had seen the films of the Zagreb case, I made arrangements to go to Liberia to do some experimental work—not on Liberians, but on chimpanzees of which Dr Bray had a good stock at the Institute of Medical Research. On my way to the airport, I went to the Central Veterinary Laboratories at Weybridge to pick up some heparinized blood containing *B. divergens*. On arrival in Africa I was quickly hustled through the customs by Dr Gunders and Dr Bray and we went straight to the laboratory to inoculate intravenously the blood into two chimpanzees, one of which we splenectomized simultaneously with the inoculation; the other had been splenectomized 3 years before. A third chimpanzee with an intact spleen was later inoculated with the parasite as a control. The two splenectomized chimpanzees developed the infection, which strangely enough appeared earlier in the animal which had been without a spleen for 3 years—a 6- instead of a 26-day incubation period and with many more parasites (12 per cent of the erythrocytes were infected). This animal became severely ill with haemoglobinuria, was prostrated and on the 11th day we went to the animal house expecting to find it dead. But instead it was quite lively—the haemolysis had ceased, the parasites began to diminish in numbers and disappeared entirely after a month—as was shown by inoculation of its blood into a splenectomized calf. The control animal showed parasites for a few minutes only after the intravenous inoculation (Garnham and Bray, 1959).

Later Dr Voller and I did similar experiments with *B. divergens* and rhesus monkeys splenectomized 1–2 years previously. The results were much the same, except that the animals developed a higher parasitaemia, though none died. These rhesus monkeys showed antibodies, which were easily detectable by the Fluorescent Antibody Test (FAT) and the titre rose considerably after a second challenge. We also tested the response of an intact rhesus which had received heavy inocula of *B. divergens* and found that the FAT remained negative. In other words, this reaction is unlikely to be of any value in detecting latent infections in human beings with spleens. In fact, all these experiments indicate that the piroplasms are unable to gain *any* foothold in a primate with an intact spleen.

We also thought it would be interesting to see if *Theileria parva*, the cause of East Coast fever in cattle, could infect splenectomized chimpanzees—but the results were negative.

Of course, for a true model, it would be necessary to expose the animals to the bites of infected ticks instead of blood and it would be worth while to carry out a few such experiments.

Piroplasmosis is one of the most important veterinary problems of the world; the elucidation of its method of transmission by Theobald Smith in 1886 (see p. 181) was the first demonstration of the role of an arthropod in the carriage of agents of disease. Yet the full details of the life cycle of these organisms have yet to be worked out.

An equally interesting and fundamental problem is the nature of the life cycle of *Toxoplasma gondii* and other parasites in the sub-class, Toxoplasmatea, such as *Sarcocystis* and *Besnoitia*. Of the whole group, most is known about *Toxoplasma gondii*, the little intracellular protozoon, first seen by Nicolle in his Tunisian gundis in 1908, and now known to infect about a third of the human population of the world, as well as a great variety of animals, including many rodents, dogs, cats, cattle, sheep, pigs, monkeys, crows, chickens, etc. Splendore described the parasite in the same year (1908) as Nicolle, from his laboratory rabbits in São Paulo; 60 years later this same colony is practically 100 per cent infected with toxoplasmosis. A single species, *T. gondii*, infects them all and readily passes from one to another, though sometimes several blind passages are required before the parasite becomes fully adapted to the new host. Nearly all infections in man are 'inapparent', to use Nicolle's own word, though Nicolle never realized that the organism which he had described was later to become recognized as one of the best examples of this phenomenon. Severe cases do occur in man, but they are rare and the disease never becomes epidemic; on the other hand, epizootics of toxoplasmosis have been reported in swine, chickens, pigeons, mink and rabbits, particularly when these animals have been living under crowded conditions in captivity.

Congenital transmission of *T. gondii* is well-known in man, and in animals such as mice, in which Beverley (1959) showed that the organism could pass through nine successive generations. Desmonts *et al.* (1965) produced conclusive evidence that the infection was acquired by infants in Paris through the consumption of uncooked mutton and there are many examples of transmission through cuts in the skin in laboratory workers or in butchers and others who handle infected animals.

The possible role of arthropod vectors of *T. gondii* has long been considered, and experiments have been made with mosquitoes, *Stomoxys*, *Triatoma*, *Rhodnius*, and *Pediculus humanis corporis* (van Thiel, 1964) always with negative results. Giroud and Legac (1952) demonstrated the existence of the organism in hard ticks (*Rhipicephalus sanguinius*), taken off dogs in West Africa, and more recently de Pastoris-Castellani and Babudieri (1968) showed that their colony of soft ticks (*Ornithodorus moubata*) in Rome was infected with *Toxoplasma*. The Italian workers claim that transmission by ticks to clean animals also took place, but these experiments have yet to be confirmed.

The existence of the well-known cystic stage in the life cycle of *T. gondii* suggests that this must have biological significance, and it was felt that none of the known methods of transmission was really more than a byway; the true cycle, involving the cyst and the production of sexual stages, had yet to be demonstrated. Therefore, when Hutchison (1965) announced that he had apparently succeeded in passing the infection through the ova of the nematode worm, *Toxocara cati*, most workers felt that here at last was the solution to the problem. His initial experiment may be briefly summarized as follows:

1. (a) A cat was infected with *Toxocara cati*.

(b) Two months later it was fed for 3 days on the carcases of 6 mice, heavily infected with cysts of *Toxoplasma gondii*.

(c) Two weeks after these feeds, faeces of the cat were collected and subjected to the zinc sulphate flotation technique in order to get the nematode ova.

(d) The ova were washed and suspended in tap water and kept at room temperature for 3 months, when they were again washed and introduced into the stomach of mice. All the mice developed toxoplasmosis.

2. (a) A control cat was chosen with no *Toxocara cati* infection.

(b) The cat was fed on infected mice as in 1(b) above.

(c) Faeces were collected and the nematode-free sediment after the flotation was collected and stored for 3 months.

(d) The sediment was fed to mice which failed to become infected.

3. The two cats were then interchanged; the first cat was freed of its nematodes with anthelminthics and the second cat was infected with *Toxocara cati*. Both were then given infected mice to eat and ova were collected as before. But now, the second cat produced ova which proved to be infective to mice and the first produced negative faeces—a reverse of the original experiment.

Hutchison's work was confirmed by Duby (1968) in England and Jacobs (1967) in the U.S.A.; they extended it further and encountered another extraordinary problem, for they showed that not only does toxoplasmosis arise in mice fed on ova obtained from the faeces of *Toxoplasma*-infected cats, but (unlike in the original results) also from the faeces of cats in which *Toxocara* is absent. The latter material was still infective after being passed through a 35 μ mesh, and storage for 1 month in water and disinfectants. In other words, *Toxoplasma* in this unknown stage is able to resist the most adverse environmental conditions; it was difficult to visualize in what form it managed to survive. One was tempted to assume that *Toxocara cati* had no role to play in transmission but Duby (1908) showed in experiments of the same nature, that larvae which were freed from the well washed shells of the ova and themselves thoroughly disinfected, were capable of inducing toxoplasmosis when inoculated into mice. It later became evident that these fallacious interpretations were due to the contamination of the nematode ova and larvae by the highly infective oocysts or sporocysts of the protozoon (see below).

A further blow was given to the nematode theory by workers in Scotland, Denmark and the United States, after the delivery of this series of Heath Clark lectures. Frenkel *et al.* (1969) and Sheffield and Melton (1969) demonstrated that one strain of *T. gondii* (sheep strain M–7741) given to *Toxocara*-negative cats, produced infective forms in the faeces 2–13 days after the infective feed. These forms were resistant to sodium hypochlorite and formalin, and survived for 3 months in water at room temperature. It was suggested by these workers that investigations should be concentrated on the gut of the vertebrate rather than on the nematode egg in order to find the origin of the infective forms.

It is known that mucosa of the cat's intestine, from one end to the other, may be infected with *Toxoplasma gondii* after experimental feeds on infected mice. It might be assumed that a rapid sporogonic cycle takes place after union of gametes with the production of resistant zygotes; the latter then escape from the mucosal cells and are shed into the lumen of the intestine. Such a process was confirmed by the experiments of Hutchison (1968) and Work and Hutchison (1969) who showed that after feeding cats with *Toxoplasma*-infected mice and collecting the faeces at 6-day intervals, small cysts (9 × 14 μ) appeared in large numbers between the sixth and twelfth day. After flotation, they were washed and suspended in water for 3 weeks; the material was then passed through a 35 μ filter (to exclude *Toxocara cati* ova) and individual cysts proved to give rise to toxoplasmosis when inoculated into mice. The cysts resisted strong chemicals and after keeping for some days, two bodies (sporocysts) developed, as in a coccidial cycle.

The final solution to the problem of the life-cycle of *Toxoplasma gondii* was revealed a year after these lectures were given. Dr William Hutchison telephoned me from his laboratory in Scotland late one Saturday night in December 1969 to say that the discovery had just been made. In the previous few weeks, the Scottish, Danish and North American workers had recognised that the cysts were in fact oocysts, whose contents were seen to divide into two bodies (sporocysts) after some days, and the sporocysts were eventually found to contain four elongated bodies (sporozoites). The inevitable, however incredible, conclusion was reached that the oocysts were coccidial in nature and apparently belonged to the genus *Isospora*. If this were so, a search of the intestine of the infected cat should reveal the schizogonic and gametogonic stages of the parasite. This search would have to be done at the right time and in the right part of the intestine, possibly within rather narrow limits. Hutchison, with the intuition which had guided him throughout the work, chose these limits and his telephone call was to tell me that he had found profuse developmental stages of the parasite in the epithelium of the small intestine. Within 10 days, a letter had appeared in the *British Medical Journal* and within a month, the full paper (Hutchison *et al.*, 1970).

Two important questions remain to be settled; how does man become infected, and what is the significance of the well-known cystic and pseudocystic forms of the parasite. The association with cats has long been recognized and it seems probable that the natural or biological cycle is initiated by the ingestion of the highly infective and resistant oocysts from the excreta of domestic animals. The part played by the long-known cyst and pseudocyst in toxoplasmosis may be entirely aberrant and might possibly be explained by the proliferation of the products of schizogony in abnormal sites. Other coccidial parasites (e.g. *Eimeria necatrix*, van Doorninck and Becker, 1957) are taken up by macrophage cells and are transported in the submucosa of the intestine to the epithelium. Possibly the

---

[1] For this work Hutchinson was awarded the Robert Knock Prize and Medal, the first time that the honour had been bestowed on a British scientist.

merozoites of rupturing schizonts of *Toxoplasma* have a similar fate, and instead of remaining in the intestine are taken to different organs (e.g. lymph glands) to initiate the proliferative phase and later to the brain and muscle for their final 'dead end' as cysts, at least in man. Such multiplication of *Toxoplasma gondii* is easy to induce in tissue cultures of macrophage and other cells (Pulfertaft *et al*, 1954). It may be noted that 'aberrant' cycles are absent in probably all other forms of coccidia, with the possible exception of the metastatic development of *E. arloingi* (Lotze *et al.*, 1964) in the lymph nodes of sheep.

A further problem is the apparent inability of any animal other than the cat to produce this new cycle, and many have been tested.

Coccidians are usually very host-specific parasites; yet *Toxoplasma* in its 'aberrant' phase has the widest variety of hosts perhaps for this very reason, or because it had learnt how to infect such a range from the time when vertebrate animals started to evolve and to come in contact with this ancient parasite.

It is thus necessary to abandon entirely the nematode theory of transmission of *Toxoplasma*; however, the life-cycle of another mysterious protozoan parasite involves nematodes. Lee (1969) has very recently demonstrated by electron microscopy, a high rate of infection of the developmental stages of *Histomonas meleagridis* in the oocytes of *Heterakis gallinae*.

Sarcocystis (including the mysterious 'M'-organism or *Frenkelia*) and *Besnoitia* belong to the same group as *Toxoplasma* and in many morphological details closely resemble each other. Their life cycles also are unknown, though there is a little evidence that a faecal route is concerned in some species of *Sarcocystis*. *Sarcocystis*, though apparently rare in man, is even more widespread than *Toxoplasma*, and material for study is available at the nearest butcher. The parasite has been known since 1843, and it really is time that its mode of transmission should be clarified.

I have mentioned a few problems whose solution seems likely to open new paths in parasitology. There are innumerable others, including a few in which I continue to be involved myself and with my students. I refer particularly to the still unexplained phenomenon of relapses in malaria, the functions of certain organelles in the Sporozoa and the transmission of the parasites of lizard malaria.

The problem of the transmission of lizard malaria has long interested parasitologists, and yet until very recently has proved insoluble. There have been several unconfirmed reports of oocysts and even sporozoites in various species of wild mosquitoes, which had fed on infected lizards, but much more substantial evidence became forthcoming early in 1970. Ayala and Lee showed that two species of sand fly (*Lutzomyia vexatrix* and *L. stewarti*) developed very large numbers of oocysts in their midguts after feeding on lizards infected with *P. mexicanum*. Motile sporozoites (4–6µ in length) appeared in the haemocoelomic cavity and in the salivary glands 11 days after the sand flies had fed. Until malaria pigment has been demonstrated in the oocyst and transmission to clean lizards has been effected, the role of these insects in lizard malaria must remain open, but the findings are most suggestive. If confirmed, this will represent the first instance of a malaria parasite (*Plasmodium*) developing in an insect other than a mosquito, and the systematic position of these common parasites of lizards may have to be revised.

There should be no difficulty in *finding* problems, either in the classical fields of parasitology itself, or, for those suitably trained, in immunology, biochemistry and genetics. A few years ago, the British Society of Parasitologists made a list of research projects for submission to the Council concerned with the programme for the International Biological Year. Unfortunately, they did not prove quite suitable for its purpose, which is directed primarily at improvements in the food supply of the growing population of the world. However, the Society's main recommendations included subjects which are of permanent interest. The problem of host-parasite specificity is always before us. What is the factor which makes the blood, but not the liver, of the chimpanzee just unsuitable for *Plasmodium falciparum* or the Guatemalan *Simulium* for the *Onchocerca* of Cameroun? Why does *Brugia malayi* develop easily in *Mansonia* but not in *Aedes aegypti*, and what is the relationship between certain species of schistosomes and their molluscan hosts? And then within a specific vector, why are only a few individual members of a population of *Glossina morsitans* susceptible to *Trypanosoma rhodesiense* or of a specially selected colony of *Aedes aegypti* to *Plasmodium gallinaceum*? Why is *Anopheles gambiae* a good, and *A. culicifacies* a bad vector of malaria parasites? Why does *Plasmodium gonderi* grow perfectly to the mature oocyst in *A. atroparvus* and perish immediately in the sporozoite stage? The factors controlling the migrations of *Leishmania* in *Phlebotomus*, of *Dirofilaria* in mosquitoes, of *Babesia* in ticks are very imperfectly known, as are the migrations of these parasites in their vertebrate hosts.

The nature of host specificity is fundamental and yet practically nothing is known about it. The superficial data are available, but the biochemical, biophysical, physiological and genetic details are largely missing.

This subject represents the microcosm of parasitology. The macrocosm is constituted by the 'natural focus' and its changed pattern after human intervention. The parasitologist has taken this subject to his heart and reflections on it comprise the content of this book. But many interesting features remain to be investigated, including the nidality of the carcinogenic viruses (of Burkitt), the demyelinating rickettsiae (of Legac), the mongolizing toxoplasm (of Jirovec), the decerebrating Ascaris (of Schulmann), and the allergic nematode (of Buckley) .

But, above all, the influence of man on his environment requires immediate attention. In this polluted atmosphere, the gloomy forebodings of Erda are only too audible in our own *Götterdämmerung!* The balance of nature is being upset: the vectors of the great killing diseases go down and the human population goes up, pesticides destroy pests but also beneficial animals, chain effects are started whose consequences we have yet to see. The long-term action of the newer synthetic drugs, insecticides, fertilizers, etc., has yet to be assessed, although so far there is little evidence of the more subtle injuries— sterility, cancer or genetic defects—which were predicted. No surveys on a large scale appear to have been made to observe the effect on the ecosystem of public health or agricultural campaigns. Such research forms the natural extension of the

Pavlovskian doctrine and could well be carried out in many parts of the tropics and subtropics where interference with the natural foci has taken place.

The blunderbuss methods of the modern age might well be replaced by a gentler approach in which the environment would not be so severely damaged. Biological control of the vectors of parasite infection is still in its infancy and although difficult and unspectacular, it offers possibilities which deserve further research. To a parasitologist, such a method has great appeal.

Even biological control however may disturb the balance of Nature in unwanted directions. The sugar plantations of Mauritius were ravaged by the cane rats, so the mongoose was introduced from India in an attempt to reduce their numbers. But the rats had kept down the population of snakes, and when the rats decreased, the snakes began to multiply and devoured the eggs of the partridge; the chief sport of the French emigrés on the island was thus ruined. A similar situation arose later in the nineteenth century in S. Thomas and Grenada in the West Indies, where the mongoose has become a serious pest, ravaging the young of domestic animals and acting as a reservoir of rabies.

Desultory attention has been paid to biological control of mosquitoes, which started many years ago with the use of larvivorous fish (*Gambusia* and *Lebistes*). These were disseminated widely and it would be interesting today to survey the long term effect in different conditions in various countries. Anyone who has watched the avidity with which mosquito larvae are devoured by *Toxorhynchites*, *Eretmapodites* and *Lutzia tigripes* has felt inclined to encourage the process, and the writer even added colchicine to his breeding bowls to obtain bigger and better megarhines. Laird (1967) seeded Nukuono atoll in the Tokelau Islands with the fungus, *Coelomomyces*, in an attempt to control *Aedes polynesiensis*, which is susceptible to this parasite, and Reynolds (1969) introduced the spores of *Plistophora culicis* to the island of Nauru in the Pacific to control *Culex fatigans*; but these measures have not yet been developed enough to obtain clear-cut results. Reynolds has recently shown that *Nosema stegomyiae* invades the testes of mosquitoes and when this parasite is introduced into a colony, the egg production becomes greatly reduced. In the field of agricultural parasitology, biological control has had substantial success, e.g. by the release of sterilized males of the screw-worm (Baumhover *et al.*, 1955), while Vankova (1958) used *Bacillus thuringiensis* for the control of insect pests in the forests of Moravia. The prickly pear was eradicated from parts of Australia by the introduction of the parasitic beetle, *Cactoblastis cactorum*.

Recent experiments on a different aspect of biological control by Biocca and Massi (1965) have shown that hydatid cysts in the organs of infected sheep can be 'biologically purified' by feeding the material to pigs and chickens which are insusceptible to the parasite, but are nourished by this otherwise dangerous offal.

Personal problems

Like most research workers, the parasitologist is confronted not only with the problems which arise directly from his work, but by certain ethical or personal problems.

The experimentalist is often faced with the problem of volunteers, and in spite of the rules for the ethics of their use issued by the Medical Research Council (1967), the ardent research worker may have to find a way around them, just as he has to overcome any other obstacle, and it is classical for him to use himself, or next best in our subject, a splenectomized chimpanzee. The rules are a useful guide, and when wisely applied by a local system in an Institute, a case can nearly always be made out to meet the requirements. It is much better to have the advice and approval of your colleagues, for in your enthusiasm for the project you may be blind to a not so obvious danger. Some years ago, I wanted to make a trial of the comparative pathogenicity of *Entamoeba histolytica* and *E. hartmanni* in a group of medical students. I consulted the late Sir Neil Hamilton Fairley, F.R.S., and although he fully realized the need and importance of this problem, he advised strongly against the investigation because there does not exist at present an absolutely certain cure for amoebiasis. I had no hesitation in following the advice of this experienced and ardent investigator.

And certainly beware of the path which leads to Nürnberg!

Apart from human experimentation, the exposure of one's staff to dangerous work has always been a problem, which sometimes is not recognized. It is doubtful if anyone can go through the parasitic life without incurring some penalty in the next world or some parasites in this. During our work on tsetse control of the Kuja River in Kenya, the European supervisors of the campaigns had to live near the river bank in an intensely malarious camp. In spite of warnings and advice, the first one refused to take proper precautions and after a year he was dead of blackwater fever. His successor was appointed, and within a few months, he too died of the same disease. These were only two of many tragedies in this evil though beautiful place. Even in the laboratory, B virus, *Toxoplasma*, the incurable *Trypanosoma cruzi*, *Echinococcus*, etc., are all hazards of which, some at least of us, are too careless.

A universal problem is the question of publication. It is good discipline to write up results of research at frequent intervals, even if the papers are not submitted for immediate publication. Expression in words streamlines your ideas and focuses points which need clarification and further experiment. I am no advocate of so many papers per person per year as a measurement of work accomplished; yet neither do I accept the laziness of publishing nothing, which deprives the scientific world of useful data. Nevertheless, I am aware of, and partly sympathize with the outlook of some workers who feel that privacy is an essential element of research, but hardly go as far as to recommend the example of the Abbé Kieffer, who destroyed his unique collection of *Culicoides* just before his death in 1920, so that nobody should possess it! He did, however, publish descriptions of the new species.

The parasitologist (*sensu strictu*) has a good range of journals in which he can publish his work. The more *specialist* journals are listed at the end of this chapter.

If variety is the spice of life, there is plenty of it in the parasitological life; the alternating hosts offer changing prospects at every turn of the life cycle, and in the outer environment, there are powerful presences which influence its course. The parasitologist should pick, here, his steps with care; he must avoid the corridors of power and enter instead the narrow paths through the jungle, or the untracked steppe and desert. Here at every turn he will find something new—'Something hidden. Go and find it. Go and look behind the ranges—something lost behind the ranges. Lost and waiting for you . . . go.' (Graham, 1927). This is the everlasting whisper heard by the explorer, but equally by the uniontist

And don't be afraid to *speculate* about your discoveries for, in the words of Conant, 'the turtle only makes progress when he sticks his neck out!'

I should like to draw attention to the best book ever written on the philosophy of research: *The Rules and Counsels on Scientific Investigation* by the Spanish histologist, Ramon y Cajal, given as a lecture to the Academy of Sciences in Madrid in 1897. His precepts and advice are as valid today as they were at the end of the last century, and range from how to interpret a natural phenomenon, to how a scientist should choose a wife.

SPECIALIST JOURNALS RELATING TO PARASITOLOGY

England: Parasitology. Annals of Tropical Medicine and Parasitology. Journal of Helminthology. Protozoology. Transactions of Royal Society of Tropical Medicine and Hygiene.

France: Annales de Parasitologie Humaine et Comparée. Protistologica. Bulletin de Société de Pathologie Exotique. Comptes Rendues des Séances de Société de Biologie.

Italy: Parassitologia. Rivista di Parassitologia. Archivio Italiano di Scienze Mediche Tropicali e di Parassitologia.

Germany: Archiv der Protistenkunde. Zeitschrift für Tropenmedecin und Parasitologie. Zentralblatt für Bakteriologie, Parasitenkunde, Infektionskrankheiten und Hygiene.

Switzerland: Acta Tropica.

Hungary: Parasitologia Hungarica.

Belgium: Annales de Société Belge de Médécine Tropicale.

Spain: Revista Iberica Parasitologia.

Czechoslovakia: Folia Parasitologia.

Poland: Acta Parasitologica Polonica. Acta Protistologica.
                 Wiadomiści Parazytologiczne.
U.S.S.R.:   Meditsinskaya Parazitologia (Moscow). Parasito-
                 logia (Leningrad).
U.S.A.:   Journal of Parasitology. Washington Journal of
                 Helminthology. Journal of Protozoology. American
                 Journal of Tropical Medicine and Hygiene.
                 Experimental Parasitology.
Brazil: Revista Brasileiro Malariologia e Doenças Tropi-
                 cais. Revista do Instituto Medicina Tropical (S.
                 Paulo).
Chile: Boletin di Parasitologia.
Venezuela: Archivos Venezuelan Medicina Tropicale.
Mexico: Revista Latinoamericana de Microbiologia y
                 Parasitologia.

REFERENCES

AYALA, S. C. and LEE, D. (1970) 'Saurian malaria: Development of
sporozoites in two species of phlebotomine sandflies', *Science* **167**, 891–2.

BAUMHOVER, A. H., GRAHAM, A. J., BETTER, B. A., HOPKINS, D. E.,
NEW, W. D., DUDLEY, F. H. and BUSHLAND, R. C. (1955) 'Screw-
worm control through use of sterilized flies', *J. econ. Entomol.* **48**, 462–6.

BEVERLEY, J. K. A. (1959) 'Congenital transmission of toxoplasmosis
through successive generations of mice' *Nature, Lond.* **183**, 1348–9.

BIOCCA, E. and MASSI, O. (1968) 'La "lotta biologica" alla echinococcosi',
Parassitologia **10**, 61–73.

BROWNING, C. H. (1967) 'Pathology in Britain in the first half of the
twentieth century, with a glance forward', *Br. med. J.* **3**, 359–62.

CASTELLANI DE P. (1969) 'Zecche del genere *Ornithodorus* portatrice di
Toxoplasma Gondii', *Parassitologia*, **11**, 73–5.

COOPER, D. K. C. (1969) 'Agricultural accidents: study of 132 patients
seen at Addenbrooke's Hospital, Cambridge, in 12 months', *Br. med. J.*
4, 1934–5.

DESMONTS, G., COUVREUR, J., ALISON, F., BAUDELOT, J., GERBEAUX,
J. and LELONG, M. (1965) 'Etude épidémiologique sur la toxoplasmose.
De l'influence de la cuisson des viandes de boucherie sur la fréquence de
l'infection humaine', *Revue fr. Etud. clin. biol.* **10**, 952–8.

VAN DOORNINCK, W. M. and BECKER, E. R. (1957) *J. Parasit.* **43**, 40–48.

DUBEY, J. P. (1968) 'Feline toxoplasmosis and its nematode transmission',
Veterin. Bull. **38**, 495–9.

FRENKEL, J. K., DUBEY, J. P. and MILLER, N. L. (1969) 'Fecal forms of
Toxoplasma gondii (protozoa): their separation from eggs of the nematode
Toxocara cati', *Science* **164**, 432–3.

GARNHAM, P. C. C. and BRAY, R. S. (1959) 'The susceptibility of the higher primates to piroplasms', *J. Protozool.* **6**, 352–5.

GARNHAM, P. C. C., DONNELLY, J., HOOGSTRAAL, H., KENNEDY, C. C. and WALTON, G. A. (1969) 'Human babesiosis in Ireland: further observations and the medical significance of this infection', *Br. med. J.* **4**, 768–70.

GIROUD, P., LEGAC, P. and GAILLARD, J. A. (1952) 'Mise en évidence de toxoplasmes sur souris inoculées avec des broyats de *Trombicula legaci* Marc André, 1950, recueillis sur *Lemniscomys barbasus striatus* et sur *Mylomys cuninghamei alberti*, capturés en Oubangui-Chari', *Bull. Soc. Path. exot.* **45**, 449–51.

GRAHAM, S. (1927) *The Gentle Art of Tramping*, London: Ernest Benn.

HUTCHISON, W. M. (1965) 'Experimental transmission of *Toxoplasma gondii*', *Nature, Lond.* **206**, 961–2.

HUTCHISON, W. M. (1968) 'The faecal transmission of *Toxoplasma gondii*', *Acta path. microbiol. scandinav.* **74**, 462–4.

HUTCHISON, W. M., DUNNACHIE, J. F., WORK, K., and SIIM, J. C. (1970) 'The Coccidian-like nature of *Toxoplasma gondii*', *Br. med. J.* **1**, 142–4.

JACOBS, L. (1967) 'Toxoplasma and Toxoplasmosis' in *Advances in Parasitology*, Vol. 5. Ed. Ben Dawes, London and New York: Academic Press.

KREBS, H. A. (1967) 'The making of a scientist' *Nature, Lond.* **215**, 1441–5.

LAIRD, M. (1967) 'A coral island experiment' *Wld Hlth Org. Chron.* **21**, 18–26.

LEE, D. L., LONG, P. L., MILLARD, B. J. and BRADLEY, J. (1969) 'The fine structure and method of feeding of the tissue parasitizing stages of *Histomonas meleagridis*', *Parasitology* **59**, 171–84.

LOTZE, J. C., SHALKOP, R. M., NEEK, R. J. and BEHIN, L. C. (1964) 'Coccidial schizonts in mesenteric lymph nodes of sheep and goats; *J. Parasit.* **50**, 205–12.

MAYR, E. (1963) *Animal Species and Evolution*, London: Oxford University Press.

MEDICAL RESEARCH COUNCIL (1962/1963) Annual Report, 21–25. 'Responsibility in investigations on human subjects'.

PORCHET-HENNERÉ, E. (1967) 'Corrélations entre le cycle du développement de la coccidie *Myriosporides amphiglenae* et celui de son hôte *Amphiglena mediterranea*. Mode de contamination', *Protistologica* **3**, 451–5.

PULVERTAFT, R. J. K., VALENTINE, J. C. and LANE, W. F. (1954) 'The behaviour of *Toxoplasma gondii* on serum–agar culture, *Parasitology*, **44**, 478–84.

RAMON Y CAJAL, S. (1954) 'Reglas y Consejos sobre investigacion cientifica', in *Obras literaries completes*, Madrid: Aguibo.

REYNOLDS, D. G. (1969) 'The biological control of *Culex fatigans* by the introduction of parasites', Thesis for the Ph.D. degree of the University of London.

SHEFFIELD, H. G. and MELTON, M. L. (1969) '*Toxoplasma gondii* transmitted through faeces in absence of *Toxocara cati* eggs', *Science* **164**, 531–2.

VAN THIEL, P. H. (1964) 'Toxoplasmosis' in *Zoonoses*, Ed. J. van der Hoeden, Elsevier: Amsterdam.

TUOMI, J. (1966) 'Studies in epidemiology of bovine tick-borne fever in Finland', *Ann. Med. exper. Biol. Fenniae* **44**, Suppl. 6.

VANKOVA, J. (1958) 'Kultivierung von *Bacillus thuringiensis* im Versuchsbetriebsmaßstab', *Trans. First Int. Congr. Insect Pathology Biol. Control*, 59–64, Prague.

WORK, K. and HUTCHISON, W. M. (1969) 'A new cystic form of *Toxoplasma gondii*', *Act. path. microbiol.* Scandinav. **75**, 191–2.

8

SOME PARASITOLOGICAL
ESTABLISHMENTS

FORTUNATELY the subject of parasitology is so wide that it can
accommodate workers in temperate zones and in the tropics, in
hospitals and public health departments, in universities and
research institutes, and in the Services; while the chemothera-
peutic industry provides facilities which do not exist elsewhere.
A career in a pharmaceutical firm gives admirable opportunities for research for the man or woman with tact and discretion,
though it would not be so suitable for someone who has a clear-
cut problem of their own to solve.

In this chapter, examples are given of places at home or
abroad, where the young parasitologist might well find a niche,
be he or she a medical, veterinary or zoological graduate. I
have chosen places where I have worked myself for longer or
shorter periods; there are many other establishments missing
from the list, particularly in Asia: the historical ones of India,
Pakistan and Malaya, the rapidly growing and very active
laboratories in Thailand and Japan, and the largely unknown
polytechnics of China. I have also omitted important centres,
particularly veterinary institutes, in Australia and South Africa.
As vestiges of a past era, the Colonial establishments of Britain,
the Pasteur Institutes of France, the laboratories of the other
Great Powers, still linger on in the tropics, some maintaining
considerable activity and contributing to the research programme of the country.

EAST AFRICA AND MIDDLE EAST

I shall start with the Division of Insect Borne Diseases of Kenya,
housed in the Medical Research Laboratories, Nairobi and
originating 45 years ago in the entomological section of the

laboratory under C. B. Symes, and in the malaria research section under myself in 1931. In 1944 I was instrumental in transforming these sections into the present division. Its primary task is to investigate outbreaks of vector-borne diseases in all parts of the territory and to suggest and, if necessary, carry out measures for their control. This work inevitably leads to problems which need solution in the central laboratory, and for much of the time, this so-called 'secondary' task occupies the staff. The latter research is often of a 'basic' while the field work is of an 'applied' nature. The staff includes at times (though fluctuating with the finances of the country) two or more medical parasitologists, a non-medical parasitologist, several entomologists, one mammalogist, and well-qualified field assistants. Visiting scientists, some almost permanent, are also present. This institute was originally under the control of the Colonial Office and since independence is part of the Medical Department of the Kenya Government. The facilities of the laboratory include good animal houses, insectaries and the usual type of modern parasitological equipment. The general bacteriological, biochemical and pathological departments of the Medical Research Laboratory itself are fully co-operative when need arises. Moreover, the Dutch Medical Research Centre is next door and includes on its staff an active parasitological group which works in close harmony with the Division.

The present programme (see Ann. Report) of the Division of Insect Borne Diseases comprises investigations and research on the following subjects: malaria, sleeping sickness, bilharziasis, plague, onchocerciasis, leishmaniasis, filariasis, leptospirosis, biology of sandflies and bird and lizard malaria. Apart from experimental control and use of new drugs, the work tends to concentrate on entomology (both bionomics and systematics) and animal reservoirs of infection. It is much to be hoped that the Kenya Government will continue to maintain its enlightened policy in regard to this organization.

Uganda possesses the renowned Virus Research Institute, situated in one of the most beautiful sites in the world on a small hill near Entebbe and overlooking Lake Victoria. Close by is the site of the original sleeping sickness 'laboratory', where in 1902, Castellani first demonstrated trypanosomes in the cerebrospinal fluid of patients suffering from this disease, and later the Trypanosomiasis Institute of Lyndhurst Duke, which became the Yellow Fever and finally the Virus Research Institute of East Africa. During the heyday of the existence of the Yellow Fever Institute, it was administered and largely staffed by the Rockefeller Foundation; an excellent field station was maintained at Bwambu in the tropical rain forest on the Congo border. Tree platforms were erected for the study of acrodendrophilic mosquitoes and of monkeys, including rhesus as sentinels to provide evidence of yellow fever infection. In this way, the elucidation of the mystery of African yellow fever was solved. The work has been extended to other virus diseases, and emphasis has continued to be placed on the entomological and animal reservoir aspects, though a diagnostic service is also an important function of the Institute. Here as in all the tropical laboratories, visitors are much welcomed. It says much for the foresight of the East African Common Services Organization that this Institute is still maintained; it is always subjected to criticism, because with the exception of the sporadic and rarely epidemic yellow fever, virus diseases make little impact on the popular mind.

The Institute for Malaria and Vector-Borne Disease Research at Amani in Tanzania also occupies an historic site where the old German agricultural station was transformed into a research station for the study of malaria. The laboratories and houses occupy an isolated site near the summit of a forested hill in the Usambara Mountains at an altitude of a thousand metres. The late Dr Bagster Wilson chose this remote place for a malaria research station and administered it as a Grand Seigneur. His laboratories were at Amani but his hunting country extended over the whole of East Africa. This institute provides scope for investigations of many kinds, and though its future is suspect, let us hope that this unique place will be preserved for malaria research.

The laboratories of NAMRU III in Cairo have undertaken important research in parasitology for nearly 20 years, under the indefatigable Harry Hoogstraal. Although this unit is a United States service establishment in a foreign country, some members of the staff are civilians and belong to many nationalities.

As an original exponent of the doctrine of the nidality of disease, Hoogstraal extends his work far from the confines of Egypt to Afghanistan, Malaya, Nepal, Turkey, Sudan, East Africa, Ethiopia, and Madagascar. The facilities of the unit include a small hospital in Cairo for clinical research, and general microbiological laboratories. Anyone working in NAMRU III has thus opportunities to make field trips for purposes of medical zoology over a wide area of the Old World Tropics. The scope of the work is to 'elucidate biological interrelationships between vertebrate hosts, arthropod parasites, and pathogenic agents in the epidemiology or natural history of human and animal diseases' (Hoogstraal, 1968).

The Department of Parasitology of Ain Shams University in Cairo has been ably built up by Professor Rifaat, and is engaged both in teaching undergraduates and postgraduates as well as in active research work by the senior staff, particularly in medical protozoology and schistosomiasis. I draw attention to this place because it offers interesting facilities for work in the field with the unusual natural foci of desert, oasis and delta.

The department of parasitology of the Hebrew University of Jerusalem had a celebrated history under the late Saul Adler, F.R.S. It included protozoology, helminthology and entomology in its sphere, as well as certain microbiological problems. Now, these sections have split up: microbiology has absorbed the major portion (under Avivah Zuckerman) and a new department of medical ecology has taken over certain subjects. A country like Israel which is in the process of making 'the desert blossom' needs to be professionally aware of the problems arising from such interference with the natural foci. The parasitologist knows these dangers and it is important that the classical organization of Adler should not be swamped by too much molecular biology and other high flights in our subject. At present a fascinating field experiment lies open for workers in Israel. At the same time, of course, the fine laboratories in the new departments will be used for research in immunology, entomology and other subjects.

Further to the East lies Persia which has made great contributions to parasitology, both in its ancient past and at the present day. An interesting evolution has taken place in parasitology, for which a Chair in the Faculty of Medicine of Teheran was created after the Second World War, and this was followed in 1952 by the establishment of the Institue of Malariology, later enlarged to include parasitology, tropical medicine and hygiene, and finally in 1965 it became the Institute of Public Health Research.

These developments were largely inspired by Dr N. Ansari, who later became the Director of the Division of Parasitology of the World Health Organization. The research of this Institute became channelled into the subjects of 'epidemiology, patho-biology, and ecology', or more shortly parasitology *sensu latu*, with special emphasis on malaria, schistosomiasis, trachoma, intestinal infections, leishmaniasis, and leptospirosis. These subjects are investigated in appropriate research stations in different parts of the country, e.g. Isfahan, Kazerun and Meshed. Following the tradition of Ansari and Mofidi, the staff of the institute spend as much time as possible in field trips from these research stations and their subunits. More sophisticated research is commencing in the Central Institute.

LATIN AMERICA

Latin America has a record in the history of parasitology second to none, starting in the early years of this century and flourishing in all the countries today. In Brazil alone, major achievements included the discovery of Chagas' disease, the elucidation of *espundia*, the isolation of one of the commonest human parasites *Toxoplasma gondii* (by Splendore in the same year as Nicolle found it in North Africa) and what is often forgotten, the first demonstration of a tissue phase in the life cycle of malaria parasites, viz., the exoerythrocytic forms of *Haemoproteus columbae* of pigeons by Aragão in 1908. In the State of São Paolo alone there are eight separate departments of parasitology. The opportunities for research in our subject are legion. Also, Brazil and other countries in Latin America have solved the racial problem, and the atmosphere is refreshing after experience elsewhere. The inhabitants of Brazil are the indigenous Indians, now in small numbers or localized in the deepest forests, but whose blood flows in the veins of much of the population, through inter-marriage in the early days; next the Portuguese; then a large group of Africans, survivors from the days of slavery; and finally numerous more recent immigrants from Japan, Germany and Italy. Whether mixed or pure these people all have equal rights—but what is more of interest is that everyone seems quite oblivious of race as such. They are quite colour-unconscious and your next-door neighbour or colleague may just as well be a Negro as a Braganza!

Emile Brumpt played a great part in establishing the close cultural link between Latin America and Europe, for he travelled and worked extensively in these territories of the New World. French was spoken by all educated people, and Brumpt was a name to conjure with. Today hardly anyone speaks French, instead the land is infiltrated with North American salesmen of one sort or another and the laboratories are full of their gigantic centrifuges and electron microscopes. The inhabitants regret the new 'civilization' but realize that it was inevitable owing to their isolation from Europe in the Second World War. The famous School of Parasitology in Santiago, Chile, was founded by Noé, the collaborator of Grassi, and delegates from this School were proud to add a memorial plate to the tomb of Grassi outside Rome at the time of his centenary in 1955.

The present Brazilian School of Parasitology has, as its grand maître. Samuel Barnsley Pessŏa, and his book on the subject is as much the Bible of Brazilian workers as Swynnerton's book on tsetse flies used to be to the entomologists of Africa. Pessŏa was responsible for much of the recent development in Brazil and he is still active in parasitology. He apologizes for his bad English, by saying that he could not speak English well because he had too much English blood to be fluent in foreign languages.

A characteristic of the Brazilian centres is their close link with the field. Flavio de Fonseca, the late Director of the Institute of Butantan, used to spend much of his time studying the snakes, lizards and birds of the tropical forests, and describing the life cycle of their parasites. The lively new University of Ribeirão Preto possesses departments in the Faculty of Medicine whose staff spend more time in the field than in their laboratories. The parasitology department of M. P. Barretto is most stimulating, as is the pathological unit of the Viennese, Fritz Köberle, who has solved at last the mystery of 'mega' in Chagas' disease (see p. 29); the Faculty deplores the premature death of Pedreiro da Freitas, who combined deep scientific insight with a passion for the control of the great endemic diseases of the country.

In Belo Horizonte, the Institute of Rural Endemic Diseases is engaged on interesting research on the immunological and entomological aspects of leishmaniasis, malaria and schistoso-miasis. This laboratory, like the Institute of Biophysics and the Institute of Oswaldo Cruz in Rio de Janeiro, are more occupied today with basic research. But Brazil offers in its vast territory, prospects for the study of parasites in nature which do not entail mere collecting but watching speciation in action. Fortunately, there are many Brazilian parasitologists (such as Leonidas and Maria Deane, Alencar, Martins) who have a talent for this work, and welcome in their midst visitors from abroad (such as the late Saul Adler, Biocca, the present British team at the Evandro Chagas Institute in Belém, and the numerous scientific expeditions to Amazonas and the Mato Grosso).

The same degree of enthusiasm and expertise prevails in Colombia, amongst the disciples of that unrecognized genius, Uribe, painter, novelist and parasitologist; Carlos San Martin (a student of the London School) was one of his pupils and his department of microbiology and parasitology works both in the Universidad del Valle at Cali and in field stations in the tropical jungles, where the pristine parasitology is studied before con-ditions are disturbed by advancing development. Another invigorating follower of Uribe was Santiago Renjifo whose energy was so tiresome to the Government (of which he became a Minister) that he left the country to take charge of the Pan American Sanitary Bureau in Rio. Soon after his arrival, this irreplaceable worker was killed in a motor accident. Two other followers of Uribe were Hernando Groot (celebrated for his work on *Trypanosoma rangeli*) who became rector of the Univer-sidad de los Andes, and Ernesto Osorno-Mesa (the co-discoverer with Boshell and Bugher of jungle yellow fever).

Parasitology in Venezuela has had the good fortune for many years to be under the guidance of Arnoldo Gabaldon, whose work on malaria has been referred to elsewhere in this book.

There is an open invitation from both Colombia and Venezuela for workers in our subject from abroad.

On the peninsula of Panama lies the Gorgas Memorial Institute of Tropical Medicine and Hygiene, recently much enlarged and now under the direction of Martin D. Young, the eminent malariologist from the National Institutes of Health. The old staff still continues to work here, Marshal Hertig, Carl Johnson and Fairchild, and there are many good newcomers. The spirit of Marston Bates, who told us to think like a mosquito, still hovers over the Gorgas Institute and particularly over the forests of Panama where the workers are still occupied with the habits of *Phlebotomus* and the sylvatic vectors of jungle yellow fever, undreamt of by their Founder. For historical, personal and parasitological reasons, the Gorgas Institute offers ideal opportunities.

Further north, Mexico City is another active centre of parasitology, whose study was initiated in 1918 by Barron, a pupil of the Institut Pasteur in Paris (see Aubert and Hernandez, 1966). The subject has been advanced in recent years by Francisco Biagi, who collected a good group of workers in his department of parasitology and microbiology in the National University; flourishing research on many aspects was carried out here and in the wards of the hospitals. Field work of great interest was done in the Yucatan and Quintana Roo by his staff and students. Regular training was undertaken in the Maya-haunted forests of this region, where Dr and Mrs Biagi proved that *Phlebotomus* (*Lutzomyia*) *flaviscutellatus* is the vector of *Leishmania mexicana* by inoculating themselves with flagellates from the gut of these sandflies and producing typical lesions (see also p. 23). This excellent team is mostly supported by external grants and many of its members work on the old German system of unpaid (but highly privileged) assistants to the Herr Professor. Unfortunately with the temporary departure of Professor Biagi, this staff has been fragmented into a department of 'human biology', but it may still be rebuilt.

The Technical University of Mexico City has well-qualified parasitologists in the persons of two Spanish refugees of great eminence—Professors Pelaez and Bolivar (pupils of Ramon y Cajal) and Perez-Reyes, the first person to transmit *Plasmodium berghei* (with *Anopheles aztecus*) in the laboratory and now trying to find the vector of lizard malaria parasites, of which so many species occur in this country, even at an altitude of 2500 m. The old Instituto de Salubridad y Enfermedades Tropicales in the past carried out fine research on parasitology firstly under Martinez-Baez in 1934, later by Mazzotti and Luiz Vargas, and now by Varela and Davalos-Mata. Further parasitological departments exist in the Institutes of Biological Sciences and of Veterinary Medicine.

UNITED STATES AND CANADA

Parasitology in various guises is carried out extensively in the United States, where opportunities of all sorts await the visiting scientist or more permanent worker. The institutes belong to the Armed Forces Services, Federal Government, Universities and private foundations. It is interesting to note that *medically* qualified parasitologists form a small minority of the staff. The United States has made notable contributions to parasitology in the tropics, either by providing special institutes, like the Liberian Foundation of Tropical Medicine and the Gorgas Memorial Laboratory, or more often by priming indigenous establishments with American staff and money. Thus the School of Hygiene of the Johns Hopkins University sends out parasitologists to work at the Calcutta School of Tropical Medicine, Tulane (associated with the names of Faust and Beaver) has a special interest in the Universidad del Valle in Cali, Kessel and the University of California remain in close touch with filariasis research in the South Pacific and the Rockefeller Foundation in schistosomiasis research in the Caribbean.

The North American establishments are so numerous that it is difficult to select typical ones. Perhaps today, the most interesting research and the greatest opportunities emanate from the National Institute of Allergy and Infectious Disease at Bethesda with its substations in Chamblee and until recently in Malaya. Malaria research here has at its disposal laboratories of biochemistry, immunology, chemotherapy and facilities for the use of human volunteers. Some of the leading figures have gone,

Don Eyles prematurely dead, Coatney 'retired' to Georgia and Martin Young transferred to Panama, but enough remain to make it a good centre for work. Nearby is the Naval American Medical Institute in which Clay Huff directed until recently the malaria research unit and a place of pilgrimage for all malariologists; to it belong the units abroad (NAMRU) e.g. in Cairo and Taiwan. In the vicinity also is the up-and-coming Department of Medical Zoology of the Walter Reed Army Medical Center, where interesting research in all aspects of malaria and schistosomiasis is undertaken under Elvio Sadun. In New York City, two focal points of contemporary interest are the departments of parasitology of the Rockefeller University under William Trager and of the New York University under Harry Most and Yoeli of *Plasmodium berghei* fame. Other valuable centres of different types include Stauber's department (with a singular absence of technicians) at Rutger's University where important research on leishmaniasis and schistosomiasis is being done; Levine's school of veterinary parasitology at Urbana; Carlton Herman's Wild Life Preservation Service, which operates throughout the country and provides scope for parasitologists wishing to study organisms in their natural hosts, and the United States Department of Agriculture at Beltsville, Maryland. The National Communicable Diseases Center in Atlanta is very active in many branches of our subject.

On the other side of the continent, marine parasitology is much in the hands of the Noble twins in the University of California. The Hooper Foundation was 'made' by Karl Meyer and is being developed along inspiring lines by Audy; many references to their outlook on the zoonoses will be found in this book. The National Primate Center, also in California, is an essential link in the chain of parasitological institutes, as it provides facilities for the study of the hosts with the nearest affinities to man, and, let us hope, for the preservation of the rarer species. The new UCI Center for 'pathobiology' has much interest for parasitologists, because it is intended to be not only a repository for type material and historical collections but a research centre for invertebrate parasitology, which was under the able direction of the late Edward Steinhaus.

Parasitology in the United States is often practised in watertight departments, but in the University of Georgia, the subject has been combined into a single faculty to the mutual advantage of the veterinary, zoological and plant parasitologists of the University and others from extramural colleges.

There are two famous academic institutes of parasitology in Canada, one at McGill University and the more recent department in the School of Hygiene of the University of Toronto. The former is entirely postgraduate and its policy was largely set by the late Director, Thomas Cameron, on the lines of host-parasite relationships including medical and veterinary problems and the effects of the parasites on invertebrate hosts. The School in Toronto is closely founded on the system of the London School of Hygiene and Tropical Medicine but possesses a more exciting field station than Winches Farm in Algonquin Park, to which all the staff repair in the summer months. Murray Fallis and his team have completed fundamental work on the life cycles of certain haemosporidian parasites of birds at this station which has led to a re-appraisal of the whole group.

There are other departments of parasitology in Canada where the subject flourishes, including Vancouver, and Guelph. The new department of biology at St John's University, Newfoundland is being planned on imaginative lines by Marshall Laird, and includes work on the parasitology of migrating birds and marine biology, and will involve expeditions to the tropics.

EUROPE

The modern study of parasitology began in Europe three centuries ago and work of the greatest interest is still being carried out here, even though the activities are on a smaller scale than the North American, and there are fewer specialized departments than in South America.

Britain's contribution to tropical medicine was recently reviewed by the writer (Garnham, 1968) who pointed out the great importance of parasitology to this subject. Applied in this way, the subject began in England towards the close of the last century at the Liverpool and London Schools of Tropical Medicine where the tradition has been steadily maintained. The staff at these two Schools probably spends about a third of its time in research, a third in teaching postgraduate students and a third in travelling abroad or in other activities. The annual course for the M.Sc. in Parasitology in London is unique in that nearly half the student's time is spent on a small research project and the training of the medical, veterinary and zoological graduates is much directed to research. The facilities of this School include a small Field Station (Winches Farm) in Hertfordshire (see p. 112). At the Imperial College of Science and Technology, the position is reversed in that the work of the Section of Parasitology is much more concentrated at the large Field Station at Silwood Park in Berkshire than at the college itself in Kensington; like most other centres, except for the London and Liverpool Schools, parasitology here has a more general bias. In Universities, the subject is studied at the Molteno Institute and the Veterinary School at Cambridge, Brunel, Exeter, Bristol, Salford and Leeds; Edinburgh, Glasgow and Aberdeen, and Aberystwyth, Bangor and Cardiff. The Medical Research Council has an interesting programme on parasitological research in Mill Hill and a large overseas unit in The Gambia under Ian McGregor; the Foreign Office continues to support parasitological research in the tropics. In addition, there are the pharmaceutical firms, particularly the Wellcome Foundation and others such as ICI, Glaxo, May and Baker, Ely Lily, etc. The Wellcome Trust has special parasitological units in the tropics, concerned in leishmaniasis research in Belém and Ethiopia, piroplasmosis research in Kenya, geographical pathology in Uganda and Nigeria, leptospirosis surveys in the Caribbean, etc.

The Continental centres of parasitology also mostly stem from their former tropical connections and each name conjures up the memory of some major discovery.

The Institut Pasteur of Paris must be the most famous biological institute in the world, and parasitology *sensu latu* still holds a premier place; moreover its 13 daughter institutes, although now largely autonomous, still have a particular function to fulfil and provide centres of work for French and other parasitologists.

The Institut de Parasitologie in the Faculté de Médécine of Paris was the home of Brumpt, who stamped his personality on this laboratory and the many students (of whom the writer was one) who worked at its benches. Parasitology then as now had almost too much to offer, and Brumpt used to return from his journeys laden with trophies, many of which he was able to investigate further, but others which he threw to his colleagues to do what they liked with; he never wished to collaborate or write joint papers, because, as he said, once he started this practice, it would mean that he had become incapable of doing independent work. This Institute still flourishes in the hands of former tropical practitioners, like Lucien Brumpt and Michel Larivière.

The National Museum of Natural History houses today the new Chair of Helminthology, staffed by Professor A. G. Chabaud and his enthusiastic disciple, Irène Landau, who fishes interesting new parasites out of Africa on every visit. There is much more than a museum atmosphere here, and the Chair accommodates more than its nominal subject, for Chabaud (1965) in his inaugural lecture made it clear that he was going to work not only on systematics, but on the biology, ecology and epidemiology of both helminthic and protozoal parasites. In the succeeding years he has followed out this programme, remembering however his proviso that the work must be modified whenever necessary by the unexpected results that turn up. Brumpt's field station of parasitology at Richelieu still exists under the charge of Dolfuss, and is a pleasant and useful retreat from the stresses of Parisian life. The Office de la Récherche Scientifique et Technique d'Outre-Mer (ORSTOM) fulfils a function not unlike that of the former British Ministry of Overseas Development and counts a number of parasitologists on its staff.

Centres of parasitology also exist in the French provinces; in relation to tropical medicine in the great ports of Bordeaux and Marseilles, on invertebrate and small mammal parasites in Rennes, on immunology in Lyons, on electron-microscopy of parasites in Lille, Montpellier and Clermond-Ferrand, to name but a few.

Belgium was the country where in the nineteenth century, van Beneden demonstrated for the first time the life cycle of tapeworms. An old centre of tropical research associated with many famous names, exists in Antwerp, at the Ecole de Médecine Tropicale. This was the home from which derived all the great parasitological research in the Belgian Congo, and even today, the School continues to be very active in our subject and sends its staff back to Africa to help in its problems.

Holland will forever be remembered by the name of Leuewenhoek, who made parasitology possible by the invention of the microscope. The Royal Tropical Institute in Amsterdam still sees the octogenarian Swellengrebel[1] on its doorstep, and possesses several flourishing stations overseas as in Nairobi and Surinam; there are also Dutch Universities in other cities, which specialize in certain aspects of parasitology like toxoplasmosis at Leyden and veterinary problems at Utrecht.

Smaller centres are to be found in the Scandinavian countries. Denmark, where last century Steenstrup demonstrated the alternation of generations in parasitic worms, is today the major focal point for toxoplasmosis research under J. Christian Siim. His laboratories in Copenhagen are in the Statens Seruminstitut which is run on the lines of a Pasteur Institute, i.e. half the time of the staff is devoted to routine and half to research. Sweden, so vital for our subject in the person of Linnaeus, still possesses zoologists who study the parasites of wild birds and small mammals (including the M organism) from the University of Lund, while the laboratory aspects of coccidiosis and toxoplasmosis are being investigated at Göthenberg and Stockholm.

Switzerland possesses the Tropical Institute of Basel, which is intimately associated with studies on the zoonoses by Geigy at its field station at Ifakara in Tanzania.

The well-known Hamburg Institute of Tropical Medicine continues to pursue the classical interest of Germany in parasitology, while at the southern extremity of the country, the mediaeval university city of Tübingen houses another tropical school and an illustrious department of protozoology, now under Grell and a century ago, under Eimer. There are also other active centres in Germany, particularly in Bonn, under Piekarsky at the Medical Faculty and Scholtyseck at the Zoological; in Munich, under Krampitz; in Hanover, under

[1] Died New Year's Day 1970.

Denigk and Friedhoff; at Hohenheim, under Frank; and in the renowned pharmaceutical establishments, such as Bayer (Gönnert and Haberkorn).

In Lisbon, at the Institute of Tropical Medicine, still works the famous and ancient Aldo Castellani, as well as the present and past Directors, Cambournac and Fraga de Azavedo. Portugal is notable in that it is one of the few countries which has erected a public monument to one of its distinguished parasitologists, França. He stands on a pedestal at Cintra and at the base crouch two guinea-pigs.

Italy has the longest history in modern parasitology, based on the researches, first of Francesco Redi (the begetter of chemotherapy—see de Carneri, 1967), Cestoni and Spallanzani, of Lancisi and Bassi and of countless others. There is an active Italian Society of Parasitologists, which, like similar Societies in the U.S.A., Mexico, Chile, Britain, France, Germany and Poland, meets in different cities annually to discuss their problems. In Rome, today as in the past, parasitology receives major attention. The Istituto Superiore de Sanitá houses several departments concerned with our subject. Protozoology, under Augusto Corradetti, is active both in the laboratory and in the field, and is directed chiefly to malaria (life cycles and immunology) and leishmaniasis. Entomology is in the experienced hands of Saccá and Bettini, while an important department deals with leptospirosis and other infections under Babudieri. The Istituto di Malariologia 'Ettore Marchiafava' closed its doors in 1968 with the departure of Raffaele, though the 'private' Field Station of the Coluzzi family at Montecelli is still maintained. The University of Rome has a vigorous department of parasitology in charge of E. Biocca, where research is conducted on helminthology, entomology and protozoology; field work on these subjects is carried out at the special station in the Maremma in Tuscany, and in the tropics, e.g. in the Amazonian forests. This Roman School has produced teachers who have started up parasitology in other Italian Universities, like Mantovani in Bologna and Babudieri in Trieste.

Elsewhere in Italy, particular places of interest to the parasitologist include the Istituto di Genetica of Cagliari, where Frizzi works on mosquito vectors of disease, and Milan, where

Ivo de Carneri and his associates are engaged on studies of the intestinal protozoa and other organisms.

Although the Eastern European countries have had no direct tropical interests, they have long been centres of parasitology.

In Moscow, the Gamaleya Institute has extensive facilities for research in all aspects of the subject, and as the home of the late Academician Pavlovsky, it is permeated with his theories of the zoonoses. A subsidiary but important function of the Institute is the provision of practical help in the control of epidemics, both by the production of vaccines and by participation in campaigns in the field. Of specific interest are the Departments of Natural Focality of Disease under Petrischcheva, and of Rickettsioses under Zdrodovsky, and the laboratory of Electron Microscopy under Avakian. The exchange of research workers from abroad is a recognized commitment of the Institute (Baroyan, 1966). The other important Institute of Parasitology in Moscow is the Martsinovsky Institute of Medical Parasitology and Tropical Medicine, whose major achievements have been in the realm of helminthology, under Academician Skrjabin, of malaria under Serguiev and Dukhanina, and of epidemiology under Moshkovsky. Many of its members have worked in the tropics and elsewhere abroad, where they have introduced new techniques of great interest as in entomology by Detinova and Dolmatova. Field work in parasitology is carried out in the subtropical southern republics. Joint research projects are in operation between this Institute and the London School of Hygiene and Tropical Medicine. In the University of Leningrad, important work on the life cycles and fine structure of *Babesia* and coccidians was carried out by the late E. M. Cheissin, whose premature death represents a great loss to our subject; his department is excellently equipped with modern apparatus and should still serve as a focal point in these subjects. The Institute of Cytology of Professor Polyansky is more concerned with free-living than parasitic protozoa, and is occupied with the physiology of all these organisms on the fundamental lines deriving from Dogiel (1964).

Czechoslovakia is celebrated for the work of Lambl and Mendel in the past, and for the flourishing Schools of parasitology which exist today. The grand-maître is Academician Otto

Jirovec, Emeritus Professor of the Charles University, whose researches and publications on the parasitic protozoa are innumerable and whose travels abroad have made him a familiar figure in all parts of the world. From his School has derived much of the staff of the Modern Institute of Parasitology of the Czechoslovak Academy of Sciences. This Institute is directed by Bohumil Rosicky, and carries out research on all the branches of our subject, natural focality, protozoology, helminthology, entomology, arbovirology and mycology. Expeditions into the field are a notable feature of the work and there are two permanent out-stations at Klec in Bohemia and Valtice in Moravia.

The Department of Insect Pathology under Jaroslav Weiser is another parasitological establishment in Prague which specializes in microsporidia and in other agents concerned in biological control. Here as elsewhere in Czechoslovakia exchange of. personnel from research laboratories has long been welcomed,

Parasitology is actively pursued today in all parts of Poland largely owing to the inspiration of Academician Stephánski and perhaps because of the expert attention the subject receives in the Faculties of Science in the various Universities. The research institutes are devoted to different aspects of the subject. At the core is the Nencky Institute of Biological Sciences of Warsaw; on the top floor is Parasitology today under Michajłow who works on the protozoan parasites of fresh-water invertebrates, but with Stephánski actively collaborating on helminthological problems; on the ground floor is the Department concerned with free-living protozoans under Dryl and a large graduate staff doing fundamental research on ciliates. Academician Raabe heads the Department of Protozoology of the University, while the more applied sides of parasitology are the function of the Institute of Hygiene (under Mme Dymowska) and the Veterinary Institute at Puławy.

Parasitology in Poland owes much to Kozar who has been President of the Parasitological Society for many years and who is engaged in important researches on immunity, trichinellosis, and with his wife on toxoplasmosis in the Veterinary School of the University of Wrocław. His work takes him from the Tatra Mountains at one end of the country to the bison-infested

Białystok Forest at the other. A big department of parasitology exists in the medical faculty of Poznán which works under Professor Gerber on classical branches of the subject. The parasitology of tropical infections is assuming some importance at the Baltic ports, where several foreign-trained specialists (Zwierz, Dzbenski and Sduzarski) conduct research on amoebiasis, trypanosomiasis, etc.

Rumania is the country where Babes first discovered piroplasms and opened the way for Theobald Smith's fundamental discoveries in the United States (see p. 181). It also was one of the first countries to use malaria therapy for cases of general paralysis; the studies on immunity which arose out of this work by Academician Ciuca (who died in 1969 at the age of 84) and his colleagues are classics of malariology. Close collaboration has been maintained between the Rumanian workers and the Malaria Reference Centre of Horton in England since the earliest days and is still in operation. Professor Lupascu is in charge of the Department of Parasitology of the Instituto di Cantacuzino, which is as flourishing an institute as any in Europe.

The veterinary school in Belgrade under the late Academician Simic made notable discoveries on leishmaniasis and toxoplasmosis, and the department of parasitology in Zagreb under Richter contributed greatly to the eradication of malaria from Jugoslavia.[§]

This chapter has provided the briefest sketch of some places in the world where parasitology is carried out. In our subject we are often required to compare the behaviour of parasites in different regions, and for this reason, as well as for broadening our outlook, exchange visits are desirable. The World Health Organization and the various national foundations frequently help in the financing of such activities, and there is usually a warm welcome everywhere; in fact there is often an urgent invitation to participate in the research or even to come on to the staff. For the young parasitologist who wishes to study problems abroad, he or she should have no difficulty in going for a year to the United States and Canada, perhaps less easily to the Continent, but certainly to many places in Africa and elsewhere in the tropics, where parasitologists are badly needed

PLATE V

FIG. 15. Alphonse Laveran.

FIG. 16. Camillo Golgi.

PLATE VI

FIG. 17. Manson *en famille* in Amoy. The children from left to right are Edith Mary Manson (the future Lady Manson-Bahr), Patrick Thorburn Manson, and David Manson.

for teaching and research. A useful summary of the opportunities for scientific interchange abroad has recently been published by the Department of Education and Science (1969).

REFERENCES

AUBERT, E. M. and HERNANDEZ, P. M. (1966) 'Microbiologia y parasitologia medica', *Prensa Med. Mexicana* **31**, 40–41.

BAROYAN, O. V. (1966) *The Gamaleya Institute for Epidemiology and Microbiology*, Moscow: Mir.

DE CARNERI, I. (1967) 'Francesco Redi and chemotherapy in parasitology', *Trans. R. Soc. trop. Med. Hyg.* **61**, 275–6.

CHABAUD, A. G. (1965) 'Leçon inaugurale du course de zoologie (Vers) prononcée le 4 Novembre, 1964', *Bull. Museum Nat. Histoire Naturelle* **37**, 87–103.

DEPARTMENT OF EDUCATION AND SCIENCE (1969) 'Scientific interchange', Science Policy and Organization Bull. no. 2.

DOGIEL, V. A. (1964) *General Parasitology*, Rev. Polyansky and Cheissin, Transl. Z. Kabata, Edinburgh: Oliver & Boyd.

GARNHAM, P. C. C. (1968) 'Britain's contribution to tropical medicine, 1868–1968', *Practitioner* **201**, 153–61.

HOOGSTRAAL, H. (1968) 'Acceptance of the Henry Baldwin Medal', *J. Parasit.* **54**, 197–9.

HOOGSTRAAL, H. (1969) 'A brief history of the NAMRU-3 medical zoology program', *J. Egypt. Public Health Assoc.* **43**. In press.

9

SOME GREAT PARASITOLOGISTS OF THE PAST

THE HISTORY of parasitology goes back to the dawn of civilization in Ancient Egypt and Paul Ghaliounghi (1963) discusses the deep insight of the Sekhmet priests into various helminthic infections, such as tapeworm, hookworm and round worms. Examination today of mummies demonstrates the existence of bilharzial ova, and probably the well-known abdominal swelling, typical of the Akhnaton family, was due to this disease.

Knowledge of parasites slowly accumulated and Hoeppli (1959 and 1969) relates the discoveries which were subsequently made in different parts of the world. The morphology and clinical effects of some worms and arthropods were scientifically studied in the seventeenth and eighteenth centuries, the poet Redi (1684) having 'Fathered' the subject in Tuscany. Then in the nineteenth century, gifted biologists arose in many countries; just to list a few, Leuckart in Germany, Steenstrup in Denmark and van Beneden in Belgium; Fedschenko in Russia and Mendel in Moravia; Pasteur in France, Cobbold in England and Leidy in the United States.

This chapter is devoted to a dozen key figures who made discoveries which still have relevance for us today, in that the final picture has yet to be drawn. They were all men who were born in the last and died in the present century. They were characterized by their obsession with the life cycles, rather than the taxonomy of parasites, and most of them followed up their observations and theories by practical measures for the control of disease. Nearly all founded schools and had ardent followers, while their genius was sometimes inherited by their descendants.

The examples are taken from eight countries, and are chosen on a national rather than a subject basis, because each race has its own particular contribution to make, an attitude to the

subject or a peculiar insight, which is as distinctive as the different schools of nationalistic music.

The chief discoveries of most of these pioneers were made in
outposts, far from academic influences; the practical difficulties
of everyday life, and the isolation together stimulated the ger-
mination of ideas. Their example is encouraging for all who
work in the tropics without the latest equipment and perhaps
without much finance. Practically all derived their inspiration
from the field, usually in wild and remote parts of the world;
few of them were city boys but were born and bred in picturesque villages in the heart of nature—Grassi near Como and Golgi
in the shadow of the Alps; Manson near the banks of the
Scottish Don and Pavlovsky of the Russian Don; Danilewsky in
the Ukraine and Chagas on a farm in Minas Gerais; Robert
Koch in the Harz Mountains and Theiler in a gentle valley in
Switzerland south of the Rhine. It is not surprising that from
their earliest age these men became attracted to natural
history, and just below the surface of nature at that time lay
unexplored the virgin field of parasitology. Today we have to
probe deeper into molecules and genes in order to follow up their
discoveries, but nevertheless have first to penetrate the upper
strata, or we should lose our way.

These short biographies try to indicate first the *soil* in which
the genius flourished or in other words, the nidality or nest of
the parasitologist; then the chief discoveries are summarized,
and finally the relevance of their work to conditions today.

ALPHONSE LAVERAN 1845–1922

The last chapter of this book should certainly begin with an
account of Laveran who himself opened a new chapter in our
subject. Not content with his discovery of the malaria parasite,
he spent the rest of his life in prolonged researches on other
blood protozoa of man and animals and parasites of all kinds.

Laveran was born in a house on the Boulevard St Michel and
learnt to walk in the nearby Jardin du Luxembourg, but he was
soon taken by his parents to North Africa. He was the son and
grandson of doctors, originally domiciled in Dunkirk. His father
had a distinguished career in military medicine and, like

Alphonse the son, he worked both as professor at the military college of Val-de-Grâce and in the army in Algeria. His mother was from Lorraine and belonged to a military family (de la Tour et Lallemand) famous in the Napoleonic era. Laveran was thus conditioned for medicine and the army, while a sojourn in Algeria in childhood gave him a taste for Africa that he never lost.

His doctorate thesis was on the regeneration of nerves and a histological background is apparent through much of his subsequent work. He entered the army and served in the Franco-Prussian war, in which Robert Koch was also involved on the opposite side on the Alsatian front.

In 1878, Laveran returned to Algeria and he immediately set about investigating malaria; he found plenty of material, including many fatal cases of cerebral malaria, and spent his time between the wards, the mortuary and the microscope in his little laboratory. The history of his discovery in Constantine of the malaria parasite has been told countless times. The date, 6 November 1880, when Laveran first recognized the exflagellating body, is perhaps the most important day for tropical medicine. In his own words, 'on examining a fresh preparation of blood taken from a soldier suffering from malaria, I observed with astonishment a series of thin and transparent filaments on the periphery of round pigmented bodies. These moved with great agility and their living nature was incontestable. I soon found similar elements in the blood of other patients suffering from malaria and I had no longer any doubt as to their parasitic nature'. Although Laveran was completely convinced that here at last was the aetiological agent of the disease, his discovery met with a sceptical if not hostile reception from the experts. The less eminent at that time were more disposed to accept it. Thus, Sternberg in Philadelphia demonstrated blood films containing malaria parasites, about which Osler was scathing; Richard, a young colleague of Laveran's in Algeria not only confirmed the existence of the parasite, but described additional stages, but Marchiafava and Celli in Rome were unsatisfied; the relatively unknown Metchnikov in 1887 took films with malaria parasites from Russia to demonstrate to the great Robert Koch in Berlin, was kept waiting outside the laboratory for some time and then had the ignominy of seeing his specimens dismissed with a scornful glance by the German (see Nuttall, 1924). But Laveran himself in 1884 had the satisfaction of convincing Pasteur, Roux and Chamberland of the reality of his organisms, by demonstrating to them exflagellation under his microscope in the laboratory of Val-de-Grâce. At that time, the sexual nature of this phenomenon was unknown, and even today there are many gaps in our knowledge about the origin of gametocytes and the precise cytological changes which occur in gametogony.

Laveran saw all three species of human *Plasmodium*, but for a long time thought that these were but varied aspects of a single polymorphic organism. He strongly objected when the Italians named the three species separately, and still more when they gave his name, *Laverania*, to only *one* of the three that he had discovered. This so-called 'unitary hypothesis' remained in favour for a long time, and there has always been a tendency, especially among Anglo-Saxon authors, to avoid, if possible, new names and species. Wenyon (1926) for instance, refused for many years to accept the validity of *P. ovale*, and many North American investigators today prefer to designate the different types of monkey malaria parasites (*P. cynomolgi*) as the M strain (after Col. Mulligan), B strain, etc., rather than *P. cynomolgi cynomolgi*, *P. cynomolgi bastianellii*, etc., while the Belgian School tends to look askance at the various subspecies of rodent plasmodia which have been recently described.

In those early days in Algeria, Laveran was faced with the problem of the identity of strains, and, with his limited range of techniques, it is natural that he could not discover everything about the parasite; much more astonishing is the accuracy of his observations as reflected in his paintings. Even today, we are still faced with the problem of the multiplicity of strains; it is one of the lacunae in our knowledge and the question is of great practical significance in regard to immunity, the possibility of vaccination for prophylaxis, in the study of genetics, and for the treatment of the patient.

Laveran did not rest on the laurels which were showered upon him. From his discovery of the plasmodia in the blood of man, he passed to a new subject—'comparative haematozoology', transferring his attention from malaria of man to malaria of birds, including pigeons, larks, sparrows, greenfinch, and later to partridges suffering from an epizootic of the disease near the river Loire in France. From birds, he went to cold-blooded animals and found a large range of protozoa in frogs, tortoises, lizards, snakes and fish. His interest spread to piroplasms of domestic animals and their transmission by ticks. Laveran had by now transformed himself into a protozoologist, and worked on numerous parasitic forms including coccidia and sarcosporidia; he was the first to demonstrate the extraordinarily powerful toxin present in *Sarcocystis*; the substance is sarcocystin, whose properties are still not properly understood. A minute dose will kill a rabbit in a second. His experiences in North Africa brought him into contact with Biskra button and other forms of leishmaniasis; his book (1917) on this disease is the first classical study of the subject.

Laveran was one generation too soon to appreciate the problem of parasitology in the holistic sense; in the same terrain in North Africa, Nicolle penetrated more deeply into the natural foci of infections, to be followed by the Sergents (1929) in Algeria and Georges Blanc in Morocco. They were all proud to serve science and their country in the territories of their Grand Maître.

On leaving the army, Laveran went to the Pasteur Institute of Paris and when he was later awarded the Nobel Prize, he gave the proceeds to the Institute to build a laboratory of Tropical Medicine. In company with Mesnil, he worked on trypanosomes, and their 400-page monograph on trypanosomiasis is still in everyday use. His last work, in 1921, was devoted to the flagellates of insects and in the latex of Euphorbia. Studies on flagellates in Euphorbia and other milk weeds have been continued, and interesting work on the bionomics of these infections is in progress now in Georgia, U.S.A., where Barclay McGhee (1969) is tracing the progress of epibiotics, initiated by the annual migration from Mexico of infected vectors (*Oncopeltus fasciatus*).

Although Laveran had the aloof reserve typical of his Flemish blood, he led a happy life, devoted to his family and to science.

CAMILLO GOLGI 1844–1926

Laveran's immense discovery of the malaria parasite needed much development before its final implications were realized; nearly twenty years were to elapse before the mosquito cycle, and nearly seventy years before the liver cycle were found. In the meantime, the Italian malariologists concentrated on the parasite in the blood, the great constellation of Grassi, Marchiafava, Bignami, Bastianelli, Celli, and Dionisi working in Rome, and a lone star, Golgi, in Pavia in Northern Italy.

Golgi was born in 1844 in the small town of Corteno, situated just off the Val Tellina, and dominated by the Adamello glaciers. His father was a provincial doctor in this wild and beautiful part of Lombardy. Golgi studied medicine at the ancient University of Pavia, of which, in after years, he became the Rector. Like Laveran and Marchiafava, he was attracted by the histology of the nervous system, and for his graduate thesis he submitted a dissertation on the anatomical and aetiological features of mental disease.

Golgi loved Pavia and its university where, a century earlier the great biologist Spallanzani had worked, and where, in his own time, Grassi was to come as a medical student.

Three years after graduation (in 1872), however, he left Pavia for a small town, Abbiategrasso, near Milan, to take charge of a hospital for incurables. The move proved of great significance, for the new appointment gave him the opportunity to experiment with staining methods for the nervous system. No obstacles were allowed to interfere with his research. He rigged up a laboratory in the kitchen of his modest lodgings and here, within a year, he had discovered the silver method of staining. This discovery revolutionized knowledge about the anatomy, physiology and pathology of the brain and, in 1875, he was called back to Pavia to occupy the Chair of General Pathology and then of Histology.

The Spanish histologist, Ramon y Cajal, was the first to modify Golgi's technique, and he made further advances in knowledge of the physiology of the brain. Both workers were awarded the Nobel prize in 1906 for their work on the fine anatomy of the nervous system; but the Spaniard totally ignored the Italian in the celebrations in Stockholm as Ramon was in disagreement with Golgi's conception of the neurone, making many attacks on what he termed Golgi's false theories on the mode of transmission of impulses in the cells of the grey matter of the brain.

In 1880 Golgi discovered the curious reticular structure in the cytoplasm of all cells, which was later to bear his name—the Golgi apparatus, or the Golgi body. We study this today with the electron microscope, but are still far from knowing its real function, except that it is probably a modified form of endoplasmic reticulum concerned in enzyme production, and is said to be the main agency for the production of large carbohydrates (Neutra and Leblond, 1969).

Marchiafava discussed the recent discovery of malaria parasites with Golgi when the latter visited Rome in 1884. In addition to the eternal enchantment of Rome, there was at that time the 'final charm bestowed by the malaria—for if you come hither in summer and stray through the glades [of the Villa Borghese] in the golden sunset, fever walks arm in arm with you and death awaits you at the end of the long vista' (Hawthorne, 1804–1864).

Golgi was intrigued by the challenge, and in the autumn of 1885 he began his observations on patients with malaria—chiefly on peasants from rice fields in the valley of the Po. He distinguished two types of fever, quartan and tertian, and showed by observations on fresh blood, that the paroxysm is strictly related to the cycle of development of the parasite in the blood, the fever commencing at the time of rupture of the mature forms. If a second brood was present, then there would be daily fever in benign tertian; a second generation in quartan gave rise to 'double quartan' and a third to 'triple quartan', the latter again being accompanied by daily fever. These phenomena are referred to as the so-called 'law of Golgi', or 'Golgi's cycle'. This and later work on malignant tertian malaria placed on a solid foundation the idea that several species were responsible for the different forms of human malaria, and Golgi's (1889) differential diagnosis is as valid today as when it was first proposed 80 years ago.

His observations on malignant tertian malaria were largely

made in Rome and their interpretation proved difficult because of the frequent irregularities of the temperature chart and the blood picture. The problem was finally solved by the discovery that this parasite spends much of its developmental cycle in the internal organs, where the different phases can be observed in consecutive spleen punctures. The latter observations were also made by Marchiafava and his collaborators, and rather violent arguments arose between Golgi and the Romans on their correct interpretation.

The retreat of *Plasmodium falciparum* from the peripheral blood to the internal organs has continued to interest investigators, and Grassi and Manson corresponded on this subject early in the new century (see p. 160). Thirty years later, the writer attempted to trace the course of development of gametocytes and schizonts of East African strains of *P. falciparum*, but found to his surprise that the parasite picture was practically identical in smears of both blood and spleen. Probably most of the early Italian work was carried out on patients suffering from the pernicious form of malignant tertian malaria, when schizonts are always found in large numbers in the internal organs. Fortunately, today, experimental models are available, in that higher and lower primates, when splenectomized, become highly susceptible to the human parasite, and then *P. falciparum* develops easily to maturity in the peripheral blood of these animals, just as it does in splenectomized man.

Camillo Golgi remained in Pavia until the end of his life. He had to abandon his Chair during the First World War (Sacerdotti, 1926) but continued to work in a tiny laboratory in the ancient botanical gardens of the city, surrounded by students and visitors from abroad. Grassi had died a year earlier (1925); they were firm friends, and in 1924, Golgi acknowledged with gratitude the help he had received from his younger colleague in a letter which enclosed reprints of his last paper.

PATRICK MANSON 1844–1922

Sir Patrick Manson, Father of Tropical Medicine, was the first person to show that parasites are transmitted by insects, and he was to dominate the European scene for a quarter of a century.

His book on tropical diseases, first published in 1898 and now in its sixteenth edition is based on parasitology, and it is probably this linkage of the two subjects which had produced so many British medical parasitologists.

Manson was born in 1844 in the village of Oldmeldrum between Aberdeen and Banff in Northern Scotland. His father was the Laird of Fingask and his mother came from the same locality. It is said (Manson-Bahr and Alcock, 1927) that Manson was essentially a naturalist; in his earliest days watching insects and frogs, and dissecting cats for tapeworm, and in old age, observing the formation of clouds, weathering of rocks, and the behaviour of birds and fish. The chief memories of his grandson, Clinton Manson-Bahr, who inherits much of his outlook, relate to fishing exploits with him on the River Don, probably

> 'Twas just whaur creeping Ury greets
> Its mountain cousin Don,
> There wander'd forth a weelfaur'd dame,
> Wha listless gazed on the bonnie stream,
> As it flirted an' play'd with a sunny beam,
> That flicker'd its bosom upon
>
> Wm. Thom.

Manson showed no signs of academic distinction either at school or at the University of Aberdeen where he began to study medicine at the age of 16, qualifying at 20 and obtaining his degree a year later.

Then began those fruitful 17 years off the China coast, first on the island of Formosa as medical officer to the Imperial Maritime Customs, and later, because of local intrigues and having been advised to leave, at the Treaty Port of Amoy on an island close to the mainland. Both these places were just outside the tropics, but tropical diseases flourished and provided Manson with all the material he needed. He was still in the Customs Service, but had a private hospital and a big medical practice which included many patients suffering from elephantiasis and chyluria. Manson worked on this disease from 1871 to 1883, and these investigations led to his greatest scientific achievements.

The cause of chyluria was already known, for the nematode, now called *Wuchereria bancrofti*, had been described in its microfilarial form in the blood by Lewis and others a year or so earlier.

Lewis was a Welshman, trained like Manson in Aberdeen, who entered the Indian Medical Service where he made many notable discoveries in microbiology.

Manson then showed, in successive stages, that the microfilaria was the aetiological agent of elephantiasis, how the adult worm in the lymphatics produced the lesions, the remarkable periodicity of the microfilaria in the blood (a subject still being investigated today see p. 98) and finally—the most important discovery of all—the role of mosquitoes in the transmission of the infection.

Manson took home leave very infrequently and spent it on furthering his research. He sometimes worked at the British Museum and, it is said, at the opposite side of a table from Karl Marx—one writing *Das Kapital* and the other 'Notes on filaria disease in Amoy'. Which was the greater benefactor of mankind?

Manson spent the next five years in private practice in Hong Kong where he established the College of Medicine and a comfortable fortune. He then returned to Great Britain and began the second phase of his career. He soon settled in London in consultant practice, but remained obsessed by tropical medicine and particularly by the significance of the transmission of infections by insects. Naturally, his thoughts turned to the most important infection of them all—malaria—a disease which had long been suspected of being of mosquito origin. Manson (1896) thought deeply about the possibilities, and after MacCallum's interpretation of the phenomenon of exflagellation of the parasite as a sexual process, realized that the midgut of a mosquito might reveal the secret, just as it had in filariasis. His collaboration with Ross on this subject provides one of the most dramatic episodes in the history of science, ending with the final proof by Ross in India of the transmission of bird malaria by mosquitoes.

The story is of great human interest in demonstrating the contrast in outlook of the two figures. Ross (undated) perhaps expressed the difference best, when he complained that his mentor took a much greater interest in the scientific than in the practical question, while he himself had not worked for the sake of *parasitology*, but in order to find a method for ameliorating malaria in the inhabitants of the tropics. The clash of personalities became severe (see p. 165) and was reflected in the attitude of the respective followers of the two protagonists. As late as 1938, there was a bitter attack by Malcolm Watson (director of the Ross Institute) on Colonel S. P. James (director of the Horton Malaria Unit and discoverer of exoerythrocytic schizogony in *P. gallinaceum*). Even today malariologists still fall into one or other of the two camps, though fortunately the animosity has lessened!

Manson was in touch with all the prominent parasitologists of Europe and America, and he had many dealings with Grassi (q.v.). He often wrote to the latter about the striking similarity in behaviour between filariasis and malaria, stating that not only were their mosquito stages 'on all fours', but that the phenomenon of periodicity in both infections was associated with the habits of the mosquito. He probably made too much of the analogy when he pointed out how the two parasites (*W. bancrofti* and *P. falciparum*) retire to the deeper organs from the peripheral circulation, and thought that both phenomena possibly had a similar biological basis, though how he could not guess.

The third great achievement of Manson was the founding of the London School of Tropical Medicine. On his return from China, he was dismayed by the little interest that existed in England about tropical medicine; both the profession and the authorities were intensely parochial in their attitude to medicine as a whole, and the subject was taught entirely on a clinical basis. Manson saw that medicine had its roots in biology, and that tropical medicine grew out of parasitology. He accordingly undertook propaganda to establish these ideas in medical education, and succeeded in 1899 in founding the London School, and in 1907 the (Royal) Society of Tropical Medicine (and Hygiene). He quickly recruited Leiper as helminthologist, Wenyon as protozoologist, and George Carmichael Low as his right-hand man. He had Castellani sent to Uganda to find the cause of sleeping sickness, Daniels to help Ross in India, and his son-in-law, Bahr, to work on filariasis in Fiji. These were only a few of the many projects which Manson stimulated.

The drive and enthusiasm of Manson inspired successive generations of young medical men in the first half of this century. They went everywhere in the tropics with a missionary zeal, accompanied by their confrères from the Continent and later from North America. As a result, the tropical lands were rendered more healthy than their homeland, and a revolution in human history passed unnoticed.

Manson died in 1922, aged 77 years; almost his last thoughts were of the London School: 'We are doing good work, both in teaching and research, and we are paying our way'.

BATTISTA GRASSI 1854–1925

Grassi was the son of Luigi Grassi who originally lived in Milan but who, soon after his marriage to Costanza Mazzucchelli, moved to Rovellasca in the Province of Como, where Battista was born. Luigi and Costanza were distant cousins; she belonged to an old Comascan family and wrote Latin verse. They had several children but the family has now died out in the male line.

The village of Rovellasca is not far from Lake Como, and Battista spent his early days in the fields and woods, collecting plants and flowers and curious stones for his father's garden. He remained all his life addicted to natural history and regretted that there were no Chairs in this subject in the Universities. He went to Pavia in 1872 for his medical training, where his fellow countryman, Golgi, had just qualified; moreover, the University was famous in the realm of biology for names like Spallanzani and Bassi.

During a vacation back in Como, Grassi was confronted with an epizootic in the local cats; he examined one that was dying and found the ova of *Ancylostoma* in its faeces and the adult worms in the upper intestine. It was a new species as confirmed by Biocca in Rome nearly a century later. On a ward round in Pavia he was able to demonstrate to his professor (Orsi) and fellow students that the cause of an incurable and undiagnosed case of anaemia was ankylostomiasis, showing them the typical ova in the woman's faeces.

Thus began Grassi's interest in parasitology, and after qualifying in medicine in 1878, to his father's dismay, he abandoned medicine and transformed himself into a zoologist, first by studying marine biology under Kleinenberg in Messina, then the anatomy of the vertebral column of fish in Heidelberg and protozoology with Otto Bütschli. Here in this Palatinate city, he met his future wife, Maria Koenen. Finally, after producing a thesis on the development of bees, Grassi was appointed in 1883 to the Chair of Zoology and Comparative Anatomy in Catania, Sicily for which his eclectic studies had thoroughly prepared him.

Grassi's activities were widespread for he was quite tireless and was gifted with a fertile imagination. He devoted seven years to the study of the flagellate protozoans in the gut of termites, for which he was awarded the Darwin Medal of the Royal Society. He equally studied the protozoa in the gut of man and demonstrated the non-pathogenicity of *Entamoeba coli* (see Corradetti, 1954); then he worked on the life-cycle of the dog tapeworm *Dipylidium caninum* in the flea, *Ctenocephalus canis*.

In collaboration with his pupil, Noë, Grassi discovered in 1900 that *Dirofilaria immitis* is transmissible to dogs by the bite of *Anopheles*. Grassi was always intrigued by the problem of why certain mosquitoes are vectors of parasites and why others are not. He would have been interested by one solution to this problem, discovered in 1968 by Coluzzi and Trabucchi, who showed that some species of mosquitoes were resistant to infections with *Dirofilaria*, because of the presence of a formidable bucco-pharyngeal armature which lacerates the microfilariae during the ingestion of the blood. Thus *Anopheles gambiae* which possesses this apparatus is insusceptible, and *A. claviger* without it, is fully susceptible to the parasite.

In 1862, *Phylloxera* was accidentally introduced into Europe from North America, and in the next twenty years had seriously damaged most of the vineyards of western and southern Europe; Grassi was not indifferent to this disaster and spent some years in studying the biology of these insects in order to suggest measures for their control.

It is often stated that Grassi's greatest contribution to science was his elucidation of the life cycle of eels and their metamorphoses in Sicilian rivers and seas.

He was occupied during much of his life on numerous entomological problems, culminating in his observations on the Italian Culicidae (see La Face, 1954), work which had been started in Italy by Ficcalbi. While in Sicily, he discovered a new genus of Arachnids which he named after his wife: *Koenenia mirabilis.*

Grassi's investigations on malaria began in 1890 in Catania, where with his colleague, Feletti, he discovered and named the malaria parasites of birds: *Haemamoeba* (=*Plasmodium*) *relictum* in sparrows and other small birds, *Haemamoeba subpraecox* in owls and *Laverania danilewskyi* (=*Haemoproteus columbae*) in pigeons. These two investigators then turned their attention to the malaria parasites of man which were not difficult to find in this unhealthy region of Sicily. They named for the first time the parasite of quartan malaria—*Haemamoeba* (=*Plasmodium*) *malariae*, the tertian parasite—*H. vivax* and the malignant tertian parasites under three designations—*Haemamoeba praecox, H. immaculata* and *Laverania malariae*. The writer possesses Grassi's own copy of this important paper, and in the section dealing with these parasites, Grassi has added in his handwriting, two significant notes on the scarcity of gametocytes of *P. malariae* and their relative abundance in *P. vivax*, points which are of prime importance in the study of the mosquito transmission of the parasites (Plate VII, Fig. 19).

Grassi moved to Rome in 1895 where he was appointed to the Chair of Comparative Anatomy in the University. His eventual quarters, in an old convent in the Via De Pretis, were mean and squalid; nevertheless, he set up the laboratory in which so much famous research on the transmission of malaria parasites was soon to begin.

The first experiments on transmission of the human parasites were attempted in Sicily in 1890 using *Culex pipiens*, but they were of course unsuccessful. Grassi returned to this research in 1897 and in August of the following year, he visited the swamps of Follonica and Fucecchio in Tuscany to study the mosquitoes in relation to malarial endemicity. In the course of this visit to northern Italy, he went from Rovellasca to Lake Como, and, from the ancient lakeside village of Bellano, he sent a postcard to his 12-year-old daughter in Heidelberg which reads as follows:

'Cara Ella, ho fatto un viaggetto di tre giorni alla ricerca di zanzare, e spero di aver scoperto quella che produce la malaria.'

('Dear Ella, I have made a three-day trip looking for mosquitoes and hope to have discovered the kind which causes malaria'.)

Grassi must have welcomed these opportunities to revisit the scenes of his youth, and like Fabrice in the Chartreuse de Parme (Stendhal) 'to have followed with his eye, all the various ridges of the sublime mountains around the lake, and the valleys that divide them, half hidden at dawn by the delicate mists that rise from the depths of the gorges'. At the same time, his thoughts must have been on the 'special mosquitoes', two species of *Culex*, and *Anopheles claviger*, and it was on the latter that his mind dwelt continuously.

Grassi returned to Rome in September 1898, burning with enthusiasm to try to infect this anopheline with malaria. Mosquitoes were accordingly collected from the vicinity of the city and were fed on patients suffering from malignant tertian malaria in the wards of the hospital of Santo Spirito; oocysts were detected in their midguts some days later. This work was done in conjunction with Bignami and Bastianelli, and in the same year, these workers had discovered the complete sporogonic cycle in *Anopheles claviger* of *P. falciparum* and *P. vivax*.

Pazzini (1935) stated in his biography, that Grassi's greatest glory in the field of malaria research was the demonstration in 1898 of the special kind of mosquito which transmits the human malaria parasites.

In 1900 appeared Grassi's most famous work *Studi di uno Zoologo Sulla Malaria*, in which the parasites are described and illustrated in beautiful detail. The monograph is dedicated to Patrick Manson, who acknowledged the tribute in a letter dated 16 July 1900 from London, in the following words: 'I am exceedingly grateful by your kindness in paying me the compliment of dedicating to me your magnificent work on malaria. ... The world knows how great the share is that belongs to you in establishing the reality of the parallelism [of transmission of filariasis and malaria]'.

Grassi and Manson were in close communication at this time; thus in 1898, Manson sent some of Ross's 'grey mosquitoes' to Grassi for identification and the latter replied that he would be happy to classify them and leave the descriptions for Ross to

PLATE VII

FIG. 18. Battista Grassi on a journey.

1°. *H. malariae* Gr. e *Fel.*

    Sin. Amoeba malariae var. quartanae (Golgi).

    Quando è piccola manda pseudopodi di solito lunghi e sottili; presto si pigmenta. Le correnti protoplasmatiche (indicate dai granuli di pigmento) sono torpide. A poco a poco l'Emameba ingrandisce, aumentando anche il pigmento, e perde la mobilità, presentandosi perciò costantemente lobosa dapprima e poi tondeggiante.

    Essa produce di solito 9-12 gimnospore.

    Invade quasi tutto il globulo rosso senza deformarne la parte periferica, che conserva il suo colore naturale.

2°. *H. vivax* (~~della~~ ~~terzana~~) Gr. e *Fel.*

    Si distingue specialmente per i movimenti ameboidi più vivaci anche nelle forme ingrandite; le correnti protoplasmatiche sono pure più vivaci. I granuli di pigmento sono più fini. Il globulo rosso, che è invaso, di solito si rigonfia e scolorisce.

    Produce l'*H. vivax* di regola 15-20 gimnospore.

FIG. 19. Page from Grassi's copy of *Parassiti Malarici* 1892, with his amendments.

Plate VIII

Fig. 20. Fritz Schaudinn.

Fig. 21. Robert Koch.

publish himself. Again in 1900 Manson wrote to ask the Italian to help Sambon and Low with their classical experiment in the mosquito-proof hut, erected in the malaria-stricken swamps of the Campagna. A little later, they arranged together the reverse experiment; laboratory-bred mosquitoes were allowed to feed on malaria patients in the Santo Spirito Hospital, and were then dispatched to London where they were reared on Manson's son, Thorburn, and a technician. Both developed malaria.

These friendly relations between Manson and Grassi may have been responsible for the rupture between Manson and Ross (see p. 159), the latter perhaps feeling that he had been let down.

A second edition of Grassi's monograph appeared in 1902 in which was added a most interesting speculation: he stated that the sporozoite must undergo a special cycle in the body of man before the blood cycle begins, i.e., in the incubation period of the disease, and in his diagram of the life cycle, places a query at this point. This idea was however quashed by Schaudinn (see p. 170) who wrongly described the entry of the sporozoites into a blood corpuscle and thereby delayed the confirmation of Grassi's theory for nearly fifty years. It is strange that Grassi let himself be misled by the invalid experiments of the wayward though famous German. But he had now become involved in the horrible controversy over priority with Ross, which was to embitter the remainder of his life.

Much has been written (e.g. Foster, 1965) about this quarrel, but today the generally accepted opinion is that Ross was the first to trace the complete development of avian malaria parasites in unidentified mosquitoes, and to state on partial evidence that the human species underwent a similar cycle. Grassi and his collaborators described the entire development of the human parasites in *Anopheles* and showed that only mosquitoes belonging to that genus were suitable vectors.

Grassi saw the Nobel Prize for this work slip through his fingers and given to Ross; today it would have been given to both. Ross had certainly made the original discovery, however amateur his approach, and confined as it largely was to bird malaria; but his subsequent campaign against Grassi until the time of his death was ungenerous.

Grassi's last seven years were spent in the malaria field station at Fiumicino where he studied the different species of *Anopheles* and speculated on the probability of different races of *A. maculipennis*, some of which carry malaria and others which do not ('anofelismo senza malaria'). He noted that the latter race bit animals and not man. His pupils and successors studied this problem for the next quarter of a century, when the genetic basis for this phenomenon was finally demonstrated.

Grassi held a clinic for the children of Fiumicino, all of whom he knew by name. He was fond of children and adored his only child, Isabella; some of his correspondence with her are preserved at the Villa Manzoni (see p. 32) and in his reply to one of her letters (1904) he chides her for the atrocious grammar and bad writing, saying that *she* complained about *his* untidy ties, but that her fault was much worse! Grassi was less at ease with older people; the bad eyesight which had threatened his work at the microscope throughout his life, probably contributed to the petulance characteristic of the man. Although a Senator, he travelled third class for long distances throughout the kingdom, or on foot, staying at the humblest *locande* and living on bread and cheese.

He died in Rome in 1925, in the arms of his daughter, who heard his last request to look after the malarious children in the village of Fiumicino. At his own request, he was buried in its cemetery, which lies in a wild and untended corner of the estates of the Princes Torlonia. At one time in Italy they used to contrast the homage bestowed on the two men—Ross, celebrated by the Institute which bears his name at the London School and with a fine memorial in the cemetery of the Holy Trinity Church, Putney, near the original Ross Institute; Grassi buried in a humble grave in an unknown graveyard on the outskirts of Rome where no monument commemorates his fame. Yet Stella (see Manson-Bahr, 1963) at the funeral oration said that 'Italians, who are not forgetful, will visit you in this lovely spot when they desire to strengthen their faith and their spirit'. Thirty years after Grassi's death, his tomb was renovated and at a special ceremony a new memorial was added by his admirers in Chile (see p. 136). More recently still, he was commemorated by the new Istituto di Anatomia Comparativa 'Battista

Grassi' of the University of Rome, and also by the addition of his name to the little school at Fiumicino on the opposite side of the street where he used to work.

In the words of his pupil Silvestri (1925), Grassi was the complete zoologist: a taxonomist who considered that systematics was the basis of all research, a physiologist who realized the fundamental relationship between the life of an animal and its environment, a biologist who appreciated the need for both pure and applied science, and who recognized the importance of studying animals in nature. In other words a paragon of our subject!

FRITZ SCHAUDINN 1871–1906

Of all the figures in these biographical sketches, Schaudinn is perhaps the most interesting; he was a genius who made many mistakes. Something must have told him that he had only a short life before him, and into this he crowded his researches, jumping to conclusions that would never have been acceptable to the more cautious investigator. His genius burnt so fiercely that it is no wonder that it occasionally misfired.

He was born in 1871 in the hamlet of Röseningken in what was then East Prussia—the territory of the Baltic Barons, and near Königsberg, the city of Immanuel Kant. This region was long inhabited by the Lithuanian Jews, and some of the best-known names in our subject today stemmed from here: the recently dead Saul Adler and Mer, and the still active Yoeli and Albert Sabin.

There are curious myths about his origin, his father is sometimes said to have been an innkeeper, from whom the son was alleged to have acquired a marked taste for wine and beer. His daughter, Elizabeth Schaudinn-Bursy, assures the writer that these stories are entirely false; her grandfather was an estate manager[1] who managed the stud farm of Schloss Werdeln, a keen horseman and a well-known figure in East Prussia. The ancestry of the Schaudinns is given by Kuhn (1949) who states that their origin in the eighteenth century was further east,

[1] Farmer and estate manager = *Landwirt* and *Gutsverwalter*—wrongly translated as innkeeper.

where the name had the Lithuanian form, 'Schaudinnes'. The mother, Dorothea Elise Zimmerman, came from a Salzburg family which had emigrated in the previous century from Austria to settle at Pillau on the Baltic coast. From her probably, Fritz derived his talent for music, for he became a good pianist, and in later life played the Mendelssohn, Schubert, Schumann and Brahms sonatas with his wife who was an accomplished violinist. After his early death, she never touched the instrument again.

Schaudinn grew up in a mournful countryside under the grey Lithuanian sky. Wiechert (1837–1950) in his novel *Die Majorin*, describes its 'dark lakes and woods, the swampy meadows and fens, where marsh rosemary, black alder, foxgloves and deadly nightshade grow; the wild swan nest here in the summer while above soar the cranes, herons and curlews'. He went to school in the neighbouring towns, first of Insterbourg and then Gumbinnen, now called Gusev (in the U.S.S.R.). He did well at school and managed to get a place at the age of 19 in the University of Berlin. He began his studies in the Faculty of Arts, but after attending a course on the new subject of protozoology he became fascinated by the protozoa and wrote his thesis on life-cycles of the Foraminifera.

Schaudinn was appointed to the staff of the Zoological Institute of Berlin in 1894. He made several trips to Norway and Spitzbergen, making observations on the Foraminifera and other free-living organisms. Then, after three years, he changed his line; he was appointed Director of a new Institute of Protozoology just outside Berlin, and started to work on coccidian parasites of various kinds; this research was probably his most important contribution to science. He began by studying the parasites of centipedes and described *Eimeria schubergi*, demonstrating the alternation of generations in the life cycle (1899). He mistook, however, the extrusions from the nucleus of the female gamete for the residue of reducing division, *prior* to fertilization.

It was later shown by Reichenow that the process of meiosis does not occur until *after* fertilization. For many years, the 'polar bodies' extruded from malaria parasites were misinterpreted in the same way, but Bano (1959) showed that meiosis took place in the first division of the nucleus in the oocyst. This work has recently been confirmed by Canning and Anwar

(1968), both in *Eimeria* spp. and in *P. gallinaceum*; in the oocysts of these parasites, the ten chromosomes of the former and the four of the latter do not divide, but cluster in half these numbers at the poles to produce the haploid numbers of five and two respectively.

Nevertheless, in other respects, Schaudinn's account of the life cycle of *Elmeria schubergi* and of further coccidians has stood the test of time. Much of this work was done in collaboration with Siedlecki (1897), and there is no doubt that Schaudinn owed much to his brilliant Polish colleague. He would have been horrified if he had lived to see the latter's tragic end. Siedlecki continued to work until the invasion of Poland on the outbreak of the Second World War; the Germans entered Cracow, and the Polish professors were ordered to report in the Aula Magna of the University. Siedlecki was late in arriving and was refused admission by the German guards. He insisted on forcing his way in, when, together with all his colleagues, he was seized and taken to a concentration camp in Germany. Here, in accordance with the policy of the destruction of the Polish intellectual élite, they were killed or died of malnutrition. Schaudinn's son, Hans, suffered a not dissimilar fate, in that during the same war he was taken prisoner in Russia and died in 1943.

Schaudinn introduced in 1899 a new nomenclature for the alternation of generations he had observed in the life cycle of the coccidia, including schizogony, schizont and merozoite; macro- and microgametocytes and gametes; and sporogony, ookinete, oocyst and sporozoite. In the following year he demonstrated very clearly the close relationship between the coccidia and the malaria parasites.

In 1901, he married Johanna Schmitt from Westphalia and spent his honeymoon in Dresden, Vienna and Graz, en route to Rovigno near Trieste where the first of his three children was born the following year. Rovigno is a beautiful old city on the Adriatic, at that time belonging to the Austro-Hungarian Empire, but now in Yugoslavia. This locality used to be intensely malarious and Schaudinn selected for his observations, the small village of San Michele, 10 km from Rovigno and at an altitude of 130 m. He journeyed there throughout the intensely hot summers of the next three years at least once a week by boat and on foot.

At the small zoological station in Rovigno, he studied the haemogregarines of the lizards and stated that they were transmitted by the tick, *Ixodes ricinus*. But his chief interest now was malaria, and he first worked on the local mosquitoes, including the vector, *Anopheles sacharovi*. His description of the anatomy and physiology of these insects is excellent. The many sufferers from malaria in the vicinity provided him with ample human material for the study of the parasite.

Rovigno, however, provided little scientific companionship, and perhaps this absence of criticism made Schaudinn careless, for certainly some of the observations that he made were incorrect.

Thus he described a process of parthenogenesis in the female gametocyte of *Plasmodium vivax*, during which the nucleus was said to divide into two, one part to disappear and the other to continue to divide, producing merozoites. David Thomson (1917) completely refuted this interpretation by showing that the phenomenon was due to the presence in a single erythrocyte of two parasites, one the uninucleate macrogametocyte and the other the multinucleate schizont.

Similarly he wrongly described an autogamous process of fertilization in the nucleus of *Entamoeba coli*, entailing the fusion of two nuclei, one of which was said to contain male and the other female chromatin.

The most serious mistake was the notorious 'Schaudinn fallacy' (1902) relating to the supposed entry of the sporozoites of *Plasmodium vivax* into erythrocytes, instead of into the parenchyma cells of the liver which, 45 years later, we (Shortt and Garnham, 1948) showed was their true destiny. The sporozoites never penetrate red blood cells.

Schaudinn (1904) was further led astray by the little owl. In the blood of this bird are two common parasites; one, a species of *Leucocytozoon*, which in Schaudinn's time was called *Plasmodium ziemanni* and the other was *Trypanosoma noctuae*. From this combination he constructed a fantastic life cycle in *Culex pipiens* which had fed on owls infected with these parasites. This cycle was alleged to involve the transformation of ookinetes of the so-called *Plasmodium* into spirochaetes, the conversion of the spirochaetes in the Malpighian tubules into trypanosomes, and the presence of flagellated gametocytes with kinetoplasts. The 'spirochaetes' were probably microgametes of *Leucocytozoon* and the 'motile gametocytes' were trypanosomes. Schaudinn continued to make observations on these parasites of the owl a few years later in Hamburg, as though he knew that there was something shaky in his earlier interpretations. It is rather surprising to find, however, that there does exist some mysterious connection between trypanosomes and spirochaetes. The presence of spirochaetes of various types in the blood often inhibits the development of trypanosomes; there is almost a cross immunity (Chapter 6, see p. 99) and both organisms exhibit mutations or survival of special strains (antigenic variants) which are responsible for the characteristic relapses. Moreover recently Ormerod and Bird (1969) have drawn attention to the remarkable similarity in their respective ultra-structure (as shown by negative staining); they possess an endoplasmic reticulum, differentiated into tubules and globules, while their pellicles are prolonged at both ends into long fibrils.

Doubtless, Schaudinn became carried away by his theoretical views and philosophical concepts (Langeron, 1907), but he never hesitated to put them to the test and bravely infected himself on several occasions with the cysts of the intestinal amoebae. He became severely ill with amoebic dysentery, but was able to demonstrate that man is susceptible to two species of *Entamoeba* —one of which, *E. coli*, is a harmless commensal and the other, which he named *E. histolytica*, is pathogenic.

Soon after these exploits, Schaudinn returned to Berlin and in the following year, 1905, made his most dramatic discovery by finding the causative organism of syphilis, *Spirochaeta pallida*. He carried out this research with Erich Hoffman in Lesser's clinic, where he examined material from primary chancres in *fresh* preparations. By this novel technique he noted that a highly motile spirochaete was constantly present; when stained films were examined the organism could only be discerned with difficulty and this characteristic led to its being named *S. pallida* or the pale spirochaete. Schaudinn showed that the organism was present in all stages of the disease, including congenital cases.

Schaudinn immediately became famous, and at that time, nobody would have dreamt of questioning his work. It has already been mentioned (p. 165) how Grassi abandoned his theories about the sporozoites when he learnt of Schaudinn's observations. But by the time of the First World War, some of his German successors, like Reichenow, began to criticize his interpretations of cytology, and in 1926, Cole in a history of protozoology, commented on the uneven quality of Schaudinn's work and that some of it was 'grossly inaccurate and speculative to a degree'. Yet, on the other hand, Schaudinn produced masterpieces based on speculations, which on such occasions came off. It was a misfortune that he did not live long enough either to enjoy his fame or to correct his errors. His portraits suggest that he was a jolly (*lebenslustig*) man.

Schaudinn tried his hand at many things. When he learnt of Roentgen's discovery of X-rays, he at once exposed various protozoa to irradiation and demonstrated the lethal effect. He foresaw the interesting research possibilities offered by this technique which has been applied in recent years to the study of genetics and cell division; unicellular organisms have proved to be excellent models, as they are relatively resistant to ionizing radiation as compared with higher animals and plants (Wichterman, 1969).

He might have done much for tropical medicine. His contact with the two most important tropical protozoa (malaria parasites and *E. histolytica*) had given him a taste for this subject, and he was appointed Director of the Department of Protozoology of the Tropical Institute of Hamburg. But his health had begun to fail and in June 1906, at the age of 34, he died of a ruptured rectal abscess, which Reichenow considered was not of amoebic origin.

Schaudinn's premature death was a great blow to the scientific world. His wife and family, including a posthumous son, were left badly off, but a public appeal from Hamburg met with great response from all over Europe, and his children received a good education. The two sons both disappeared in the U.S.S.R. during the Second World War; the daughter, Elizabeth, entered the Public Health Service as a social worker but suffered considerable privations during the wars and in the Nazi regime; she resides now in East Germany. There are twelve grandchildren alive in West Germany today, who inherit more the musical than the scientific talents of their illustrious forebear.

ROBERT KOCH 1843–1910

Koch was born at Clausthal, a village in the Harz Mountains where his father was manager of a mine. From an early age he became interested in natural history, palaeontology and geology; he was fond of chess and played the piano. He must have known intimately the hills around his home, and inevitably have climbed the highest peak, 'The Brocken', of Walpurgis Nacht fame and celebrated by Heinrich Heine (1799–1856) in the following lines:

'Very wonderful is our first view from the Brocken; each side of our nature receives new impressions, and these separate impressions, mostly distinct, nay contradictory, produce on us a powerful effect, though we cannot as yet analyse or understand it. If we succeed in grasping the conception which underlies this state of feeling, we recognize the character of the mountain. Its character is wholly German in its weakness no less than its strength. The Brocken is a German. With German thoroughness he shows us clearly and plainly as in a giant panorama the hundreds of cities, towns and villages and all around the hills, forests, and plains, stretching away to the distant horizon. But this very distinctness gives everything the sharp definition and clear colouring of a local chart; there is nowhere a really beautiful landscape for the eye to rest on. This is just our way. Thanks to the conscientious exactitude with which we are bent on giving every single fact, we German compilers never think about the form that will best represent any particular fact. The mountain, too, has something of German calmness, intelligence, and tolerance, just because it can command such a wide, clear view of things.'

Or, on a stormy evening perhaps another dimension shimmered into place and the boy became aware of the presence on the slopes above him of a more fantastic scene:

Through the desolate abyss
Sweeping the wreck-strewn precipice,
The raging storm-blasts howl and hiss!
Aloft strange voices dost thou hear?
Distant now and now more near?
Hark! the mountain ridge along,
Streameth a raving magic song!
Now to the Brocken the witches hie,
The stubble is ycllow, the corn is green:
Thither the gathering legions fly,
And sitting aloft is Sir Urian seen:
O'er stick and o'er stone they go whirling along,
Witches and he-goats, a motly throng.

Goethe's *Faust*, Part 1

Heine emphasizes the pastoral side of the German character; the Gothic element of Faust is equally fundamental and both are represented in the two German workers included in these biographical sketches.

Robert Koch was one of a family of 11 children, yet, in due course, his father managed to send him to the nearby University of Göttingen to read medicine. His doctorate thesis comprised a study of nerve cells in the uterus and the production of succinic acid. Then, for a short time, he worked as an unpaid assistant in the Department of Pathological Anatomy. After a few years in poverty-stricken general practice, broken by military service in the Franco-Prussian war, he obtained a district medical appointment at Wolsztyn, near Poznan, in what was then called Silesia. Here his genius began to blossom; anthrax was common in the local sheep and Koch started to study this disease. He bred mice in his house and used them as experimental animals, apparently without serious objection by his wife, who put up with the mice and other inconveniences for the next 20 years; but in 1893 she had had enough and divorced him on account of his notorious liaison with a young Berlin actress (see Lagrange, 1938).

He demonstrated the bacillus of anthrax and proved that it caused the disease. His 'proof' became known as 'Koch's postulates', which we are inclined to take for granted today as part of our biological ABC but, at that time, the three laws were revolutionary: the presence of the parasite in all the lesions, its isolation and maintenance in pure culture, and its ability to reproduce the disease in laboratory animals. Incidentally, the second and third of these postulates can scarcely be said to have been fulfilled for the malaria parasites: the latter can hardly be maintained in pure culture and they will only infect splenecto-mized marmosets, gibbons or chimpanzees—which are not exactly laboratory animals

While still at Wolsztyn, Koch wrote his monograph on the aetiology of infective disease and quickly became famous. He was called to Wrocław (Breslau) where he demonstrated his anthrax bacilli to Cohnheim and to a number of visitors to the latter's laboratory, including names famous in our subject: Ehrlich, Weigert and William Welch, the author of the name, *Plasmodium falciparum*. Koch discovered the tubercle bacillus and the cholera vibrio, and was awarded the Nobel Prize in 1905.

Koch paid many visits to the tropics; in 1883 he went to India and Egypt to investigate cholera, in 1896 to South Africa to study rinderpest, and back to Bombay the following year for work on bubonic plague. Altogether he spent nearly 7 years in East Africa. He studied East Coast fever and demonstrated the tissue forms of the causative organism—Koch's blue bodies. This discovery must be the first mention of a tissue phase in the life cycle of a blood parasite. He worked on tick-borne relapsing fever on the caravan routes in central Tanzania, and spent many months in the Sesse Islands and on the shores of Lake Victoria studying sleeping sickness (see Clyde, 1967).

The discoveries regarding the malaria parasite and its trans-mission by mosquitoes eventually awoke Koch's interest. At first he was highly sceptical (see p. 152); but he eventually realized his mistake and approached the subject of malaria with enthusiasm. He even added, in 1899, another name—*Plas-modium tropica*—to the long list of synonyms which designate the parasites of malignant tertian malaria. He came near to the solution of the problem of transmission of malaria when told by the Africans of the Usambaras that they 'caught malaria by being stung by mosquitoes'. Koch was the father of bacteriology (Bulloch, 1938) but was less conversant with parasitology (*sensu strictu*) and so, unlike Manson, Ross or Grassi, did not follow up this clue.

Koch visited Java and New Guinea at the turn of the century and was one of the first to appreciate the nature of hyperendemic malaria in the inhabitants. Christophers also studied the problem about the same time; a little later came Sinton, Mulligan and Covell in India, and Bagster Wilson in East Africa. Like these later investigators, Koch was impressed by the contrast between the severe reactions of the Europeans and the mild reactions of the indigenous races. Although the indigenous children were all infected, by the time they grew up they were free of symptoms; they had become immune in youth. The European children never became immunized because the administration of quinine had spoilt their chances of acquiring immunity. Koch thought that immunization would theoretically be the best way to protect the population, but dismissed the idea as fraught with too many dangers. Today, we are being forced to reconsider the possibilities of vaccination.

Koch worked in the Medical Laboratory of Dar-es-Salaam where a brass plate commemorates his researches on malaria. In pursuit of his general zoological interests, he had various wild animals brought to the laboratory so that he could examine their blood for parasites. Large numbers of baboons and grivets (*Cercopithecus aethiops*) arrived, and he quickly found malaria parasites in their blood. Koch thought that the infections resembled human malaria, but he left the description of the organism to his colleague, Dr H. Kossel, of the Institute for Infectious Diseases in Berlin, to whom he sent the slides and some infected animals. Kossel published an account of these simian parasites in 1899, without naming them, so Laveran seized the opportunity of honouring Koch and formally identified them as *Haemamoeba kochi* in the same year.

For the next half century, the parasites of the lower monkeys of Africa were regarded as belonging to the true malaria parasites, under the name of *Plasmodium kochi* or varieties thereof. In 1947, however, the writer pointed out that the infection in the blood was entirely confined to gametocytes and that the organism could not, therefore, be called a *Plasmodium*. At the same time he demonstrated the existence of schizogony of the parasite in the parenchymal cells of the liver, but it took a further 14 years before the astonishing life cycle of Koch's parasite (now known as *Hepatocystis kochi*) was discovered in *Culicoides* (Garnham *et al.*, 1961).

Robert Koch had a hard life; he possessed an arrogant spirit which led to quarrels and he was intensely jealous of his contemporaries, and particularly of Pasteur. But his travels in the tropics in the last years of his life were, no doubt, both a solace and a novelty. He died, prematurely, of coronary thrombosis at the age of 67.

ARNOLD THEILER 1867–1936

Veterinary parasitology includes the names of many distinguished men who have advanced our subject. The aim of this book has been particularly directed to the zoonoses or diseases acquired from animals, so the importance of veterinary science at once becomes manifest. Switzerland has produced two famous examples of the profession, one still alive and active at the age of 85 (Karl Meyer) and the other, Arnold Theiler, who died rather young. Both were born in the neighbourhood of Basel, graduated from Zürich, worked in the Transvaal and had equally combative and vigorous personalities.

Theiler's birthplace was the village of Frick, in the sombre though pastoral environment of a wide valley leading to the Rhine. This valley is in the heart of Europe and resounded to the passage of streams of soldiers from Italy, Hungary, Poland and Austria, Spain and France in the campaigns of the Emperors for two hundred years (1618–1801). The blood of these men saturated the soil and probably survives today in the local people who may be aware of their history, but are quite unconscious that they inhabit the birthplace of Theiler. No monument exists and no record of the house where he was born a century ago.

His father taught biology at the local school and from an early age the son was instructed in botany, and natural history amongst the beechwoods of the Kornberg or along the acacia-shaded banks of the Aar and the Rhine. After leaving school, he went for preliminary courses in veterinary medicine to the University of Berne and he took his final diploma in the subject in Zürich in 1889. He spent two years as a lieutenant in the

Veterinary Corps and then departed for some undisclosed reason to the Transvaal. The explanation, however, was probably quite simple: a craving for adventure. He entered private practice in Pretoria and had to work on a farm to earn enough money to live on; but in doing so, he had a serious accident and lost his left hand.

Theiler's abilities were soon recognized by the South African authorities and he took charge of the campaign against rinderpest which swept like a raging fire through the country in 1896; he confessed later that this was the most difficult task of his life. Before he was really able to start his scientific career, the Boer War broke out and he was commissioned in the Transvaal Artillery on the side of the rebels.

Soon after the war, Theiler was given a new veterinary laboratory at Onderstepoort, and this place became the Mecca of the veterinary world. His influence was so powerful that Onderstepoort has remained so ever since.

For the next 20 years, he was in Government Service, and during this time he made observations of great interest in parasitology. He was particularly drawn to the study of diseases caused by viruses and protozoa. His work on the piroplasms of domestic animals disclosed that there are two separate species in cattle, which were later to bear his name, *Theileria parva* (the cause of the deadly East Coast fever), and the harmless cosmopolitan *Theileria mutans*. He then discovered the tick vectors, studied their biology and indicated methods for their control. Theiler identified anaplasmosis, another important disease of cattle and showed that it also was contracted from the bite of hard ticks.

It is strange that the life cycle and systematic position of the piroplasms (including *Theileria* spp.) are still incompletely known. For instance, the supposition that sexual union of gametes occurs in the tick remains unproved, although the recent observations of Friedhoff and Scholtyseck (1968) in Germany by electron microscopy indicate that later stages of the parasite in the tick have all or most of the characteristic organelles of Sporozoa.

Theiler carried out fundamental research on the transmission of various helminthic parasites; he worked out the life cycle of

Filaria gallinarum in the chick, *Trichostrongylus* in the ostrich, etc. He was an excellent pathologist and made great contributions to our knowledge of glanders, osteomalacia and other bone diseases, acute liver atrophy (the cause of 'staggers' in horses) and various types of poisoning.

He maintained a constant interest in virus diseases and worked on horse sickness, blue tongue of sheep, heart water of cattle and of course rinderpest. He discovered methods of immunis ation against these conditions, and was delighted when his son, Max, became illustrious in the same subject.

Like many directors of departments, Theiler was accused of overspending and it was suggested to him that it would be more economical to co-ordinate his work with that of other institutes in order to avoid duplication; his reply should be nailed to the notice board in the office of all administrators; it reads as follows: 'the department should be outside general arrangements by reason of the nature and quality of its work and that the psychology of scientific workers should not have to give way to the machine-like mind of the authorities' (quoted by du Toit and Jackson, 1936). He must have been a thorn in their flesh, because he had a gruff, overbearing personality and even his own family were said to have been afraid of him.

In 1927 Sir Arnold Theiler, on reaching the compulsory retirement age of 60, left South Africa and settled down in Lucerne. The family had never severed its link with the home country and often returned for holidays; especially during the winter for skiing in Engelberg. Theiler stayed in Lucerne until 1933. 'Stayed' is not quite the right word, for during much of this time, he was frequently abroad; he was in such demand at international meetings and in foreign universities that his scientific work (except for a few observations on calcium and phosphorus deficiency in cattle) came practically to an end. This is a warning to many of us! The Theilers later made their home in London where Arnold died after a coronary thrombosis at the age of 69.

Parasitology runs in the blood of two of his children; Gertrud is a renowned acarologist who lives in South Africa and Max is the eminent virologist at the Rockefeller Institute, who received the Nobel Prize for his work on yellow fever in 1951. The protozoologist remembers Max Theiler as much for his fascinating studies of the haematozoa of the animals of Liberia.

Theiler perceived all around him in Africa the need for better agricultural production; his activities were therefore channelled to help this along, primarily by discovering the cause of the mortality of domestic animals, and then by finding methods to prevent the diseases and thus augmenting the supply of meat for the growing human population. He foresaw the future.

THEOBALD SMITH 1859–1934

This book is largely concerned with vector-borne infections and it is therefore appropriate to discuss the man who is usually said to have been the first to have discovered the transmission of infections by arthropods; certainly Theobald Smith was the first to have provided conclusive experimental proof that infection takes place by injection of the parasite through the proboscis of the arthropod (Wenyon, 1935).

He was born in 1859 in Albany, New York State, the son of German immigrants to the U.S.A. In the previous decade, Albany became the home of Henry James, but there is no record that the novelist and scientist ever met. Theobald Smith later came to know well the nephew of the novelist, who was the vice-president of the Rockefeller Institute. Not much is known of Theobald Smith's early years, but like all the figures that we are considering, he had a talent for natural history no doubt inspired by the rural surroundings, which at that time were close to the city and which led him later in life perhaps to his veterinary researches on farms. At the age of ten, he went to Germany to see the country where his father had been an itinerant tailor, and an essay survives which embodies his childish impressions of an ancient castle near Limburg in Hesse. He graduated in medicine from Albany Medical School in 1883. He was attracted to the subject of microbiology, and his German origin allowed him to read the great volume of work on this subject that was then pouring out of Germany. He was a talented organist and was particularly fond of music by Beethoven and Schubert ('his greatest pleasure was music').

PLATE IX

FIG. 22. Arnold Theiler.

FIG. 23. Theobald Smith.
(Portrait by Adolphe Borie,
courtesy of Rockefeller
University.)

PLATE X

THE ROCKEFELLER INSTITUTE
FOR MEDICAL RESEARCH
DEPARTMENT OF ANIMAL PATHOLOGY
PRINCETON, N. J.

In general a fact is worth more than theories in the long run. The theory stimulates but the fact builds. The former in due time is replaced by one better but the fact remains and becomes fertile. The fertility of a discovery is perhaps the surest measure of its survival value. What is one man's meat is another's poison in research as in other vocations. Temperament goes far towards deciding our course. In the three different environments in which I have spent my active life I have always taken up the problems that lay spread out before me in the new environment, chiefly because of the easy accessibility of material without which research cannot go on. For in the early years material and resources were exceedingly scant and this meagreness determined the direction and scope of all research. My interest in a problem usually lagged when certain results could be clearly formulated or practically applied. To continue and analyze still further every link in the established chain either failed to hold my interest or was made difficult or impossible for causes lying outside the problem. As I look back it is precisely these links that have provided innumerable problems to others. Each link has grown into a chain and the end of successive chain making is not in sight.

Sincerely yours

Theobald Smith

FIG. 24. Quotation from a letter of Theobald Smith written shortly before his death.

Eleven years after qualification, Theobald Smith became Professor of Comparative Pathology at Harvard, selecting the Chair which was so popular in those days amongst our famous parasitologists. This appointment was attached to the Bureau of Animal Husbandry which suited him admirably, for as Howard Brown points out in an obituary (1935), Theobald Smith thought it was sounder to study infectious natural to animals, rather than *human* infections in experimental animals.

Theobald Smith made many important discoveries in microbiology, including tuberculosis (distinguishing between the human and bovine races of the organism), swine fever and hog cholera, the general bacteriology of water and infectious diseases, immunology and the phenomenon of anaphylaxis. His contributions to parasitology comprise his most original research.

In close collaboration with Salmon (of *Salmonella* fame), Curtice, and Kilborne, Theobald Smith began in 1886 his investigations on red water (Texas) fever of cattle. This work illustrates the integrity of the man, and the caution of the scientist, for he would not accept any results until he had confirmed them from many angles. The brief details of the experiments on tick transmission of piroplasms are as follows:

Tick-infected cows from the southern States where the disease was prevalent, were brought to Maryland and were divided into two groups: (A) with ticks present, and (B) with ticks removed. These two groups of animals were placed in separate fields where they were allowed to run with clean (northern) cattle. The clean animals developed red water fever in Field A (with ticks) but remained uninfected in Field B (without ticks). Next, the latter animals from Field B were taken to Field A where they soon developed the infection. Later, ticks were collected from an enzootic area in North Carolina and introduced into Field C, where a herd of clean northern animals had been placed; these became infected in due course. Finally, larvae were hatched from the eggs of infected ticks, and were placed on the ears of clean northern cows, which later developed red water fever. Congenital transmission of a protozoon through the egg of an arthropod was thus proved, as well as the role of the tick (*Boophilus annulatus*) in infecting the cow by its bite.

The Rumanian investigator, Babes, had described the organism in 1888, which was later to bear his name (*Babesia bovis*), but Theobald Smith and his co-workers studied the American species in detail and named it *Pyrosoma bigemina* in 1893, though cautiously refraining from identifying it as the certain cause of the disease.

Of almost equal interest to Theobald Smith's experiments on red water fever is his work on blackhead. In 1895 he found that this important disease of turkeys and chickens was caused by a protozoon which he named *Amoeba* (now *Histomonas*) *meleagridis*. Twenty-five years later he returned to this mysterious parasite—known to be an amoeba with a flagellate phase in its life cycle—and demonstrated that embryonated eggs of the nematode, *Heterakis papillosa*, collected from birds recovered from blackhead, would transmit the infection, after ingestion by clean turkeys and chicks. As Hall (1935) states, in an otherwise rather critical review of Theobald Smith's work, this is the first (and at that time unique) record of the transmission of a protozoan disease by a helminth. This was by no means the end of the story, for Lee (1969) produced clear evidence by electron microscopy of the developmental stages of the parasite in the oocytes of female worms (see p. 121)

During the work on turkeys, Theobald Smith discovered a species of *Leucocytozoon* (*L. smithi*), of interest because of its pathogenicity to these birds and because the life-cycle of this parasite in *Simulium* was later described.

Theobald Smith did some interesting experiments on the transmission of *Sarcocystis muris* by feeding parasitized muscle to clean mice and then demonstrating that the mice later developed the infection. Transmission of the parasite was maintained by continuous passage in mice for seven years. This work was rather uncritically discussed, and several points require re-investigation. Were the controls adequate, for natural infections in some mouse colonies are frequent? Were identifiable infective forms found in the faeces and were they unaffected by disinfectants? How long did the faeces remain infective? What was the nature of the early stages of the parasite in the newly infected mice? These questions are all of direct relevance to the still unsolved problem of the life cycle of the allied parasites in the so-called subclass Toxoplasmatea (the type genus of which has now been removed to the Coccidia).

All biographers, including Claude Dolman, emphasize his sense of thrift, which some called meanness. Even in the laboratory, Theobald Smith tended to make do with two or three mice for an experiment; he said that if these were properly dissected, they would provide more information than the two or three hundred required by the statistician, which would never be subjected to the same careful scrutiny.

Theobald Smith died in 1934, aged 75, of heart disease in the New York Hospital. Shortly before his death he wrote a philosophical letter to his friend Dr Krumbhaar, which ends thus (Plate X, Fig. 24):

In general, a fact is worth more than theories in the long run. The theory stimulates but the fact builds. The former in due time is replaced by one better but the fact remains and becomes fertile. The fertility of a discovery is perhaps the surest measure of its survival value. What is one man's meat is another's poison in research as in other vocations. Temperament goes far towards deciding our course. In the three different environments in which I have spent my active life I have always taken up the problems that lay spread out before me in the new environment, chiefly because of the easy accessibility of material without which research cannot go on, for in the early years material and resources were exceedingly scant and this meagerness determined the direction and scope of all research. My interest in a problem usually lagged when certain results could be clearly formulated or practically applied. To continue and analize still further every link of the established chain either failed to hold my interest or was made difficult or impossible for causes lying outside the problem. As I look back it is precisely these links that have provided innumerable problems to others. Each link has grown into a chain and the end of successive chain making is not in sight.

Sincerely yours,

Theobald Smith.

Some of the 'links' which were missing in his day, are still gaps today, like the nature of piroplasms and the life cycle of *Sarcocystis*. Accessibility of material is referred to at the beginning of Chapter 7, and although there will never be any dearth of problems ready to hand, it is important to remember that unless one acts quickly some material will become permanently inaccessible owing to the threat of extinction of the host. In such cases, it is justifiable to mount, for instance, expeditions to Sarawak to study *Plasmodium pitheci* in the orang utan, or to Madagascar for investigations on the malaria of the lemurs.

CHARLES NICOLLE 1866–1936

Parasitology found nearly as fertile a soil in France as in Italy, and there is a wealth of examples from which to make a selection of its famous men; Brumpt, Langeron and Lavier at the Faculté de Médecine of Paris; Roux, Roubaud and Calmette at the Pasteur Institute; Georges Blanc and the Sergents from North Africa which also provided Laveran (see above); Blanchard, Mesnil, and the greatest of them all, Louis Pasteur. But perhaps the most typical of all the French workers is Charles Nicolle, who represents a key figure in the present study of parasitology.

Nicolle was born in 1866 in the city of Rouen in Normandy. His parentage was largely Norman on both sides, though from his mother he had Italian blood, a great uncle having been Harlequin in the Comédie Italienne in Paris. His father was both a medical practitioner and an expert naturalist. The son studied at the Lycée, and was a brilliant scholar, coming first in all subjects, except mathematics in which he was invariably bottom. Although he was more attracted to the arts than science his father like the father of Hector Berlioz, compelled him to enter medicine. He acquiesced more readily than the composer and graduated at the Medical School of Rouen. In 1893 he obtained his doctorate in Paris with a thesis on Ducrey's bacillus, and then spent an unhappy eight years in Rouen in practice and as a bacteriologist. Deafness struck him during this time and when he was offered the appointment of Director of the new Pasteur Institute of Tunis, he gladly accepted it, for he had already spent a congenial few years at the Institute in Paris under Roux and Metchnikov.

Accompanied by his wife and two children, Nicolle left Rouen towards the end of 1902 for Tunis, where he soon acquired new quarters for the Institute near the Belvedere Park. Here he was to stay for the next 33 years in the most eastern and earliest of the famous institutes which acted as sentinels for French parasitology along the North African littoral. The others were in Constantine (under A. F. X. Henry), Algiers (under the Sergent brothers) and Casablanca (under Georges Blanc). Only in Tunis is the name of the illustrious director still recorded with pride; in the others the inscriptions and tablets have been erased by their successors, like those of Akhnaton in Luxor by the priests of Amon.

Nicolle travelled extensively in Tunisia, through the *bled* to the deserts and oases of the South, and along the shores of the Mediterranean from the Île de Djerba (the island of the lotus-eaters) to the peninsula of Carthage, where no doubt he meditated on Flaubert's poetical lines:

'Sur la pâleur de l'aube, et tout autour de la péninsule carthaginoise une ceinture d'écume blanche oscillait, tandis que la mer couleur d'émeraude semblait comme figée dans la fraîcheur du matin.' (In the pallor of dawn a belt of white foam shimmered around the Carthaginian coast and the emerald sea seemed as if frozen in the freshness of the morning). This panorama was a great contrast to the Norman coast, with which both Nicolle and Flaubert were equally familiar.

In the early part of the century, Tunisia provided fertile ground for discoveries in parasitology; the varieties of landscapes were each typical of an important infection and everything that Nicolle investigated proved fruitful. The intense humidity of the summer in Tunis quenched the Muse of Gide, but the climate had no effect on the scientific genius of Nicolle.

In the hospitals of Tunis, he saw many cases of epidemic relapsing fever and showed that it is only the *crushed* louse which is concerned in its transmission and that its bite cannot infect. The accidental nature of this method suggested to him that it represented an incomplete and possibly recent adaptation of the spirochaete to man. He therefore sought for a more ancient association and eventually showed that the original form of relapsing fever occurs in rodents, is transmitted by ticks (*O. erraticus*) and is due to a different spirochaete (*B. hispanica*).

Nicolle worked for many years on leishmaniasis; he first discovered a simple method for growing the organism in the so-called NNN medium (Novy, McNeal and Nicolle). He demonstrated that the dog was the reservoir of Mediterranean kala-azar caused by *Leishmania infantum*; at that time (1908) the zoonosis concept was quite novel and it was Nicolle's logic and intuition which enabled him to arrive at the solution of this problem.

In his journeys in the south of Tunisia, Nicolle saw many cases of cutaneous leishmaniasis, particularly in the oascs, where the disease is known as Gafsa Sore. In the springs of Tozeur and in the neighbouring oasis of Nefta, Nicolle studied schistosomiasis and its vector, the *Bulinus* snail. This country is studded with Graeco-Roman remains, but it must also have been penetrated by the Ancient Egyptians, and the old name Nefta is surely more likely to denote the town of the Egyptian goddess Nephthys, sister of Isis, rather than to be derived from the humdrum 'naphtha' because of the presence of oil.

When Nicolle (1908) found the small protozoon, which he later named *Toxoplasma gondii*, in the liver and spleen of a gundi in his animal house, he did not realize the importance of this discovery which, in the last 30 years, has occupied the attention of hundreds of parasitologists, with the production of 8000 papers, and today *Toxoplasma* still presents one of the most challenging problems in our subject.

Giroud (the last pupil of Nicolle) in an appraisal of his master's contributions (1962) mentions how Chatton and Blanc (of all his pupils the 'most like himself') collected gundis from the south of Tunisia and that it was not until after 17 days' residence in the laboratory that the infection appeared; Giroud then suggests that the stress of captivity activated the infection, which is normally 'inapparent' in these animals.

From the time of his arrival in Tunis, Nicolle became fascinated by the problem of exanthematic typhus, that bane of armies in the field, and populations in distress. The disease was rife in Tunisia in 1903 and its mode of transmission was unknown. Nicolle was struck by the fact that although typhus was excessively contagious outside hospital, once the patient was admitted, the infection failed to spread to the neighbouring patients, nurses or attendants; Why? Then he realized that on admission, the patient was stripped of his clothing, washed and shaved, and that the agent of transmission must be on this material; it could only be the louse. In Yoeli's (1967) dramatic words, 'Here was the solution. There could be no other way, but an ectoparasite —no other vector but the louse, the ordinary louse, the grey companion of the poor.' Nicolle infected monkeys with blood from typhus patients and allowed lice to feed on them; he then re-fed these lice on a chimpanzee (possibly the first time this animal had been used in experimental medicine), which after an incubation period developed the severe disease. The problem was solved, and within a few years, the application of delousing measures had freed Tunis and much of the surrounding countries from typhus; while during the First World War similar measures prevented epidemics of the disease amongst the armies on the western front, but not on the eastern, where they were never applied. For these discoveries Nicolle was awarded the Nobel Prize in Physiology and Medicine in 1928. In the course of these researches he contracted typhus himself.

In 1931 Nicolle went to Mexico with his Polish pupil, Hélène Sparrow, and studied murine typhus, flea- instead of louse-borne—and with an animal reservoir. Nicolle suspected an evolutionary pattern in the typhus (rickettsial) group of organisms (see p. 36), and this idea was strengthened by the later discoveries of Georges Blanc, of a tick-borne variety of the disease in Morocco.

Nicolle worked also on trachoma, virus diseases such as dengue and influenza, and brucellosis (the gravity of which he recognized for the first time). He prepared vaccines and antisera at the Institute, and maintained a complete laboratory service for the city of Tunis and the country. His work for the community was greatly appreciated, and his every whim was obeyed. Although his hearing was getting steadily worse, he could still hear the band playing outside the Institute; the noise sometimes got on his nerves and he used to send a message to the bandmaster to move off. The band went!

Nicolle's works are written in clear, beautiful language which makes them delightful to read. Apart from his scientific papers, his books such as the *Destin des Maladies Infectieuses* (1933) and *Biologie de l'Invention* (1932) are full of useful advice to the research worker. A maxim which is coloured both by his Latin esprit and scientific probity runs as follows:

'La recherche scientifique est en quelque sorte, un jeu. Tout jeu a ses règles. Il ne s'agit point d'arriver, par quelque procédé que ce soit, au but le premier. Il s'agit de remplir les conditions de jeu.' (Scientific research is in some ways a game. Every game has its rules. It is not a question of obtaining a result by *any* method. It is necessary to comply with the regulations of the game).

Perhaps not everyone would accept this advice, but nobody would disagree with another saying of Nicolle's about the conduct of research: 'as long as you take imagination as your guide and realize that it is full of errors, you may have a beautiful voyage'.

Nicolle wrote several novels and fantasies, including the stories of 'Marmouse', a faun who went to sleep in Roman times and woke up in present-day France. They are scarcely more readable, though better written than Ross's efforts in this direction, such as the latter's mediaeval romance, *The Revels of Orsera* (1920). The French writer, Georges Duhamel, had the greatest admiration and friendship for Nicolle; he even gave a lecture in 1960 at Vichy on 'Charles Nicolle and silent infections'. It is strange that Ross's champion and faithful friend should also have been a writer and poet, John Masefield, who dedicated to him the poem 'in memory of a great discovery', 60 years after 'Mosquito Day', 20 August 1897. The poem begins as follows:

> Once, on this August day an exiled man
> Striving to read the hieroglyphics spelled
> By changing speckles upon glass, beheld
> Secrets hidden since the world began

Nicolle was essentially a biologist; he stated that it was useless to approach the study of infectious disease by mathematics, physics or chemistry, because vital phenomena are constantly changing, 'we only see a fragment and not the origin, a fragment which changes its form within our hands and which represents only a moment in time'. It should be clear to the

reader that the writer of this book has long been under the spell
of Nicolle.

In 1932, Nicolle was appointed to the Chair of Medicine at
the Collège de France, but he only accepted the post on the
understanding that he could spend part of his time in Tunis. His
heart had now begun to trouble him, and in the summer of 1935
he left Europe to return to Tunis where he died a few months
later at the age of 69. By his wish he was buried in the entrance
hall of his Institute, where a stone on the floor commemorates its
famous director. On this stone flowers were laid daily for years
by his devoted pupil and last collaborator, Hélène Sparrow.

Nicolle's son, Pierre, carries on the family tradition by direc-
ting a department in the Pasteur Institute of Paris, while the
daughter is a paediatrician. His brother, Maurice, was a
distinguished biologist with a similar outlook.

CARLOS CHAGAS 1879–1934

The most important malady of South America is probably
Chagas' Disease, at least this is the opinion of the inhabitants of
that continent who take an almost jealous pride in the condition
which causes several million cases a year. In the opinion of
Adler (1959) Darwin contracted Chagas' Disease during the
voyage of the *Beagle*, when he landed on the coast of Argentina
and was bitten by the 'vinchucas' (the triatomid bugs which
convey the infection); many years later he developed Chagasian
complications which Adler thought were responsible for his
death, though Woodruff (1961) and others do not accept this
view. Darwin however had been repeatedly exposed to the bites
of infected bugs during his journeys by land through the heart
of the continent, where the infection is still more prevalent than
on the coast of southern Argentina and the danger of infection
with *Trypanosoma cruzi* even more probable.

The discoverer of the disease was Carlos Chagas, born on a
remote farm in Minas Gerais in Brazil in 1879. The family had
been established there since the beginning of the eighteenth
century and his mother came from the neighbouring town of
Oliveira. Carlos spent his early years on the farm (a coffee
plantation); his father died when the boy was only four years

old. At the age of seven, he was sent to a Jesuit College in the neighbouring State of São Paulo; he ran away, was caught by the Fathers and then expelled. He was happier at his next school which was nearer home and where he came under the influence of Father Sacramento, an ardent naturalist. Thus began his interest in the natural sciences, in which however he was not allowed to indulge immediately, for his mother wanted him to be an engineer and forced him to attend the famous School of Mines at Ouro Preto. He hated the work, became ill and returned home. Fortunately he had a sympathetic uncle, a surgeon, who persuaded his mother to allow him to go to Rio to study medicine. She finally consented and he entered the Medical School of that city in 1896.

Chagas had a brilliant career as a student, working so hard that he was nicknamed 'Two Candles' as he used two candles every night studying at his books. He was also called 'The Diplomat' because he was such a persuasive speaker. He became interested in tropical medicine while still an undergraduate and worked in the laboratory of Fajardo, on malaria and other fevers, producing a thesis on 'Haematological Studies in Malaria'.

After qualification in 1901, Chagas entered private practice in Rio, but was engaged for periods on antimalarial work at the new port of Santos. Then in 1907, Oswaldo Cruz persuaded him to join the Institute at Manguinhões, situated at that time in country surroundings, outside Rio. Chagas continued to work on malaria and mosquitoes, and became an excellent entomologist; his name has been given to a subgenus (*Chagasia*) and species (*Anopheles chagasi*) of Brazilian mosquitoes.

His interest in malaria control made him realize the importance of the *domestic* vector: the adult female *Anopheles* living in the house is the villainess of the piece. This idea was only taken up much later by James in India and Park Ross in South Africa, and today is the basic principle of malaria eradication. Another interesting discovery of Chagas was the presence of much quartan malaria in Central Brazil (near the River Acre), which he suspected might be due to a new species, owing to the greater virulence, and especially because of the curious oedema which accompanied the infections (E. Chagas, 1936). It was not until many years later that *P. malariae* was definitely proved to be responsible for this nephrotic syndrome by Giglioli (1930) further north in British Guiana, Carothers (1934) in Kenya and by other African workers, as summarized by Edington and Gilles (1968).

The great discovery of Chagas was made in the course of an anti-malarial campaign in the 'backlands' of Minas Gerais, where the Central Brazilian Railway was in course of construction along the valley of the Rio das Velhas. An epidemic of malaria was raging in the inhabitants, but there were also many other diseases of unknown nature. Chagas stayed in this region for more than a year and became aware of the common insect— the '*barbeiro*' (the '*vinchuca*' of Darwin's acquaintance in Argentina) which infested most of the human habitations (see p. 29).

The country is primitive enough today; in Chagas' time, it consisted of the fierce backlands described so vividly by the Brazilian novelist Guimarães Rosa (1908–1969) in *The Devil to Pay in the Backlands*. The scene is laid in the sertão and its wide empty spaces in northern Minas Gerais. 'Go there. Some things you will still find. Here are the sources of many great rivers: the Carinhanha and the Piratinga. They rise in marshes, enormous groves of buriti palms. There the anaconda loops and coils. The thick kind that throw themselves upon a deer and wrap themselves around it, crush it—thirty handspans long! All around is a sticky mud, that holds fast even the hooves of mules, pulls the shoes off one by one. In fear of the mother snake, you see many animals waiting prudently for the time they can come and drink, keeping hidden behind clumps of palm. The sassafras trees provide shelter around the pool, and give off a good smell. The alligator roars once, twice, three times, a hoarse roar. The alligator lies in wait—bulging eyes, wrinkled with mud, looking evilly at you. On the ponds not a single winged thing alights, because of the hunger of the alligator and the saw-toothed piranha. The hills become more heavily wooded as you approach the headwaters. A wild bull may dash out of the scrub, enraged because it has never seen people before . . . it's worse than a jaguar. We saw such large flocks of macaws in the sky that they resembled a blue or red cloth spread out on the back of the hot wind.'

Chagas was much interested in the *barbeiros*, or assassin bugs (*Triatoma infestans*) and on dissection of the hind guts he found numerous flagellate parasites (*Crithidia*). Infected bugs were then fed on clean marmosets (*Hapale penicillata*), in whose blood, three weeks later, a new species of trypanosome of very striking morphology was demonstrated. Chagas named it (1909) *T. cruzi* in honour of his much admired Master, Oswaldo Cruz, and later the same year placed it in a new genus *Schizotrypanum* on account of its peculiar developmental cycle in the vertebrate. Even at this time, he recognized the potential importance of the parasite and feverishly sought the natural vertebrate host. He returned to the infected houses, and took blood from the cats. Here at once he saw the same characteristic trypanosome and within a day or two, he found the identical organism in the blood of a sick child. This little child, then aged 3 years, is still alive and well today. She is lucky, because *T. cruzi* produces a great variety of symptoms, of which the most dangerous relate to the heart (see p. 29). Chagas studied the human disease in detail and though some of the conditions may have been incorrectly ascribed to infection with the trypanosome (idiocy, paralysis, and goitre), he was right in emphasizing its importance as a clinical entity. On the site of the discovery, at Lassance, at a solemn ceremony, the condition was named Chagas' Disease. The Chagasian aetiology of other obscure complaints has now been demonstrated (Köberle, 1968), particularly the conditions of 'mega'—megacolon, megaoesophagus (*mal de engasgo*) etc.

The true mechanism of infection through the bite of the bug was demonstrated a few years later by Brumpt (1912) while on one of his prolonged visits to Latin America. Brumpt showed that the metacyclic trypanosomes are not inoculated by the proboscis of the insect, but that, during the process of feeding, the bug defaecates and the organisms pass out posteriorly in the faeces and contaminate the wound made by the bug, or a scratch in the vicinity, or the eye, as the result of rubbing by the finger of the half-asleep victim.

Chagas had the same biological outlook which characterizes all our men; in a lecture on tropical medicine in 1925, he emphasized how the abundant and varied flora and fauna of Brazil influenced the prevalence of parasitic and particularly protozoal diseases. He showed, for instance, that Chagas' Disease is a zoonosis and incriminated armadillos and monkeys as feral reservoirs (de Magalhães, 1944).

Chagas had two sons of whom the elder closely followed in his father's footsteps both in the laboratory and in the 'backlands'; he was killed, aged only 35, in an aeroplane disaster on the way to visit his little daughter Tatiana in boarding school in São Paulo (Baccllar, 1963) The memory of this son is enshrined in the Instituto Evandro Chagas in Belém, where several of our own students are carrying on the work that Chagas, père et fils, began. The younger son, Carlos, is a brilliant biophysicist, who at the age of 27 was appointed to the Chair of Physical Biology in the Institute of Biophysics in Rio and is now his country's representative to UNESCO in Paris.

Chagas had a warm and generous personality. He never refrained from acknowledging his debt to the people who helped him in his work, including the engineer, Mota, who first showed him the '*barbeiro*'. He published only 48 papers, and never added his name to the numerous communications which emanated from the Institute of Oswaldo Cruz, unless he was directly associated with the work. It is an example to follow.

BASIL DANILEWSKY 1852–1939

While Laveran was feverishly searching for the cause of malaria in a remote town in North Africa, Danilewsky was about to begin his studies on bird malaria in the Ukraine, and during a long life made many contributions to our subject.

Danilewsky was born at Kharkov, the son of a clock maker who had his own little workshop where he invented improved actions for watches. There were three brothers, Basil, Alexander and Constantine, all of whom eventually became physiologists. Basil studied medicine in the University of Kharkov and obtained his doctorate in 1877 with a thesis entitled 'Investigations on the physiology of the brain'. A senior student in Kharkov at that time was Metchnikov, and both were inspired by the enthusiasm of their teacher, Shchelkov, to study *comparative* physiology. Metchnikov became interested in the problem of immunity in parasitic infections and eventually demonstrated the role of phagocytes in diseases of all living creatures from water fleas to man. Danilewsky became diverted to the subject of malaria: he thought that unicellular blood-inhabiting organisms would act as simple models for the more complex physiology of vertebrates, and he found plentiful material in the vicinity of his *dacha* (villa) on the outskirts of Kharkov, in the form of the haemocytozoa of birds, reptiles and amphibia.

Metchnikov fulfilled his destiny in Paris, but Danilewsky found all that he needed in the Ukraine: 'How intoxicating, how magnificent is a summer day in the Ukraine! How luxuriously warm it is when midday glitters in stillness and sultry heat, and the blue expanse of sky, arching like a voluptuous cupola, seems to be slumbering, bathed in languor, clasping the fair earth and holding it close in its ethereal embrace! In it, not a cloud; below, not a sound. Everything seems dead; only in the airy heights above a lark is trilling and the silvery notes tinkle down upon the adoring earth, and from time to time the cry of a gull or the ringing voice of a quail resounds in the steppe. The towering oaks stand, lazy and carefree, like aimless wayfarers, and the dazzling gleams of sunshine light up picturesque masses of leaves, casting on to others a shadow black as night, but flecked with gold when the wind blows. Like sparks of emerald, topaz and ruby the insects of the air flit about the gay kitchen gardens with their stately sunflowers'. This is how Gogol (1809–1852) describes the landscape as seen by a traveller coming westwards to the 'Fair at Sorotchintsi'; Danilewsky took blood films from the larks and quail, and examined the insects, but the rest of the environment had to wait a few years, for a Pavlovsky, before it, too, became transformed into the scientific picture of landscape epidemiology.

Danilewsky's contributions to physiology included the discovery of trypsin, the chemistry of muscular contraction and the existence in the brain of a centre regulating the activity of the heart. In his studies on the physiology of the brain, he also made observations on the phenomenon of hypnotism and hypnotized experimental animals, an approach which has never been followed up and still offers a clue or possible method for the elucidation of this mysterious process. Another interesting pointer to the future was Danilewsky's experiment on the attenuation of snake venom by a high frequency current and the subsequent use of the material for vaccination. One immediately thinks of the recent research on irradiation and attenuation of *Dictyocaulus* larvae and the highly successful vaccination of calves against this nematode, and the still newer work of Corradetti *et al.* (1966) who exposed trophozoites and sporozoites of malaria parasites to X-radiation and found that the parasites then acted as a perfect prophylactic vaccine against the infection. Pautrizel *et al.* (1966) adopted a rather different approach; they exposed the vertebrate host instead of the parasite to radiation —or rather to a strong magnetic field—the parasites were no longer able to kill the host and the strongest immunity developed.

In 1878, the University sent Danilewsky to Germany to work under Karl Ludwig, the well-known physiologist in Leipzig, but, after this journey abroad, he rarely left the Ukraine. He held various posts in the University of Kharkov or in research institutes in the city (Hoare, 1939; Finkelstein, 1955). He became Professor of Zoology, Physiology, and Histology in the Faculty of Science in 1883, and Professor of Physiology in the Faculty of Medicine three years later.

Danilewsky held liberal if not radical political views, which at once brought him into conflict with his reactionary colleagues. This hindered his academic career, but left him free for his research and, after the revolution, he was immediately appointed Director of the Endocrinology and the Organotherapy Institute, a position which he still held in extreme old age and which enabled him to work on hormonal physiology. As late as 1931 he wrote a paper on the influence of insulin on the sympathetic nervous system; his last publications were in 1938, the year before his death.

Danilewsky mentioned that he was quite unaware of Laveran's discoveries when, in the summer of 1884, he himself started to examine the Haemosporidia of birds and cold-blood animals. Danilewsky in fact created this name for a new order of the Sporozoa. He described and correctly identified trypanosomes, haemogregarines and microfilariae, and made a particular study of the malaria-like parasites of the red and white blood cells. He (1889) observed the phenomenon of exflagellation of gametocytes (*Polymitus*), and even pointed out the superficial resemblance of the flagella to spermatozoa. This was a decade before MacCallum was to demonstrate the sexual nature of the process.

Danilewsky was convinced that exflagellation was a unique phenomenon, only occurring in these Haemosporidia and quite absent in the free-living protozoa. He observed the process most often in owls, rollers and shrikes and said that it began a few minutes after the blood was taken. Exflagellation of *Plasmodium*, however, does not start until at least 10 minutes have elapsed; in *Haemoproteus* the process begins much more quickly, within a few minutes and, for this reason, probably many of Danilewsky's observations relate to *Haemoproteus*. This is confirmed by his pictures, and also by the fact that *Haemoproteus* rather than *Plasmodium* occurs principally in the type of birds he was examining; in the blood of owls he clearly differentiated the non-pigmented large gametocytes of *Leucocytozoon*.

There are so many curious bodies in the blood films of these animals that it is not surprising that Danilewsky was confused in some of his interpretations, just as we are today still in doubt about the exact nature of some of the bodies found in the blood of snakes and lizards; as Gordon Ball (1965) said at the Second International Congress of Protozoology in London, there are many structures of uncertain nature to be found in the blood of reptiles, including artefacts, haemogregarines, haemosporidians, piroplasms, and even viruses, which have been placed in at least 17 genera of most dubious validity. Nevertheless, by 1886, Danilewsky clearly recognized the similarity between his avian Haemosporidia and the human malaria parasites as described by Laveran. He strongly shared the latter's opinion that only a single species of parasite was concerned (Danilewsky, 1896). Thus he did not contribute any names to the taxonomic jigsaw puzzle which was about to arise in the nomenclature of these organisms.

It is difficult to recognize the species of malaria parasites which Danilewsky encountered in his birds, though they undoubtedly included *P. relictum*, *Haemoproteus columbae* and *Leucocytozoon sacharovi*. He described pigmented parasites in the blood of the hoopoe, which may well have been the first observations on the species of *Plasmodium* which bears the writer's

PLATE XI

FIG. 25. Charles Nicolle as a young man.

FIG. 26. Carlos Chagas as a young man in academic dress. (Portrait by Marques Junior.)

PLATE XII

FIG. 27. Basil Danilewsky.

FIG. 28. Academician Pavlovsky, Sir Harold Himsworth and Professor Irina Bihovskaya at the London School of Hygiene and Tropical Medicine, 1963.

name. He noticed that the parasites of other birds were present in young erythrocytes and were particularly numerous in the bone marrow; such observations suggest that this parasite was *P. elongatum*. The birds which were most commonly infected around Kharkov were rollers, finches, shrikes, thrushes, buzzards and owls.

Danilewsky worked on these parasites for 12 years (1884 to 1896) and then completely abandoned the subject, although, like Laveran, he was well aware of the significance of the study of bird malaria in the interpretation of human malaria. He actually stated that work on the former would facilitate experimental studies on malaria in general. Fifty years later he was still alive and capable of appreciating the truth of this prophecy by seeing the brilliant research which then began on avian malaria: the discovery of the new synthetic drugs by Kikuth in Germany and the demonstration of the 'third cycle' of the parasite by Raffaele in Italy, Huff in the U.S.A. and James and Tate in England.

It is interesting to note that a devoted worshipper at the shrine of Danilewsky, Reginald Hewitt, spent about the same time working on bird malaria (1937–1949) and then forsook the subject, fortunately not before producing his own valuable monograph.

EUGENE NIKANOROVITCH PAVLOVSKY 1884–1965

In Chapter 2, the natural focus is shown to be the primary general principle in the study of the zoonoses, a concept which was first developed by Pavlovsky. Much of the modern approach to our subject depends upon his doctrines. It is appropriate therefore to conclude this book with a brief description of the Russian scientist and of his work. In his last years, he was an imposing figure in the uniform of a Lieutenant General of the Soviet Army and with rows of ribbons on his breast. He was equally imposing in his younger days, on horseback in Turkmenia, leading his team of parasitologists, including Petrishcheva across the desert.

Abroad he was nearly always accompanied by his daughter, Irina Bihovskaya, the eminent helminthologist of the Zoologica,

Institute of Leningrad. He depended on her for communication with foreigners, for latterly he was stone-deaf and understood only his native tongue. In spite of the aura of power and prestige which surrounded him, he was both approachable and convivial at those gay and uninhibited evening parties in which the Slavs so much delight.

Pavlovsky was born in 1884, the son of a school-master, at Birjutch near the River Don in the Voronezh province of Russia (Hoare, 1965). After matriculation in 1903 at the Borisoglebsk Gymnasium, he entered the Military Medical Academy of St Petersburg where he graduated in medicine in 1909 and obtained the M.Sc. in zoology and comparative anatomy in 1917. His teachers included the renowned physiologist, Pavlov, the histologist Maximov (who emigrated to Chicago and died of home-sickness) and the zoologist Cholodkovsky. He worked, particularly with the last, on the anatomy of insects and wrote a practical manual on methods for their dissection and study. His doctorate thesis was on the poison glands of arthropods and fishes.

Pavlovsky was appointed to the Chair of Biology and Zoology at the Military Medical Academy in 1921 and held this post for the next 40 years, during which it became changed into the Chair of Parasitology. In fact, in 1933, he established the famous Department of Parasitology at the Institute of Experimental Medicine in Moscow, and eventually into the Department of Natural Foci of Disease at the Gamaleya Institute (see p. 146) where today his portrait, medals, films and records are as carefully preserved and revered as those of Manson in London.

After the Revolution, Pavlovsky applied his immense theoretical knowledge and tireless energy to the problem of the arthropod-borne infectious diseases in all parts of the Soviet Union. He and his staff made more than 160 expeditions on horseback, lorry or on foot through the taiga, tundra and steppes of the immense region in Central Asia, the Far East and Siberia. Gretchaninov's haunting song 'The Dreary Steppe' must surely have run in his head during these journeyings, with its syncopated accompaniment, echoing the footsteps of the plodding wanderer:

Dreary the steppe where I'm journeying,
Never a flower to be seen
Never a tree where the nightingale
Sings in a bower of green.

Or, he may have had on his mind the symphonic poem 'In the Steppes of Central Asia' by Borodin, who had also been a student at the Military Medical Academy but half a century earlier. Pavlovsky himself was something of a poet and wrote a book on *Poetry and Science in Russia* (1958).

From these experiences, Pavlovsky synthesized a theory on natural foci of disease, for which his original training in medicine, zoology and biology had thoroughly prepared him, allowing his mind to focus on all the factors concerned. A large vocabulary has arisen in connection with landscape epidemiology, and altogether there are more than 50 terms, including 17 relating to the nidus or focus of infection. These expressions are usefully defined in a glossary by Norman Levine in the English translation of Pavlovsky's last book on *Nidality of Disease* (1966).

This constituted the third and last phase of Academician Pavlovsky's career which dates from 1938 when he first wrote about landscape epidemiology in relation to hydatid disease and infections with *Diphyllobothrium latum*. He extended his observations to microbiology, showing the interwoven pattern of events and the importance of multiple infections. He repeatedly emphasized the need for establishing the complete ecologo-parasitic situation in new localities.

Pavlovsky worked on anopheline mosquitoes and malaria as early as 1924, and must have been proud of the successful eradication of this disease from the Soviet Union after the Second World War, while the eradication of cutaneous leishmaniasis from some regions in the southern republics was directly due to his efforts. As Pavlovsky grew older, he turned his attention increasingly to the control or eradication of these infections. He had a great devotion to his country and its inhabitants, and felt that it was his duty to introduce modern ideas of hygiene into its remotest corner. His last years were clouded by some personal tragedies, and he died in Leningrad in 1965, at the age of 81.

REFERENCES

ADLER, S. (1959) 'Darwin's illness', *Nature, Lond.* **184**, 1102–3.

BACELLAR, R. C. (1963) *Brazil's contribution to tropical medicine and malaria*, Rio de Janeiro.

BALL, G. H. (1965) 'Some protozoa and other bodies found in the blood of reptiles and amphibians', *Progress in Protozoology*, 127–8. Amsterdam: Excerpta Medica.

BANO, L. (1959) 'A cytological study of the early oocysts of seven species of *Plasmodium* and the occurrence of post-zygotic meiosis', *Parasitology* **49**, 559–85.

BROWN, J. H. (1935) 'Theobald Smith Obituary', *J. Bact.* **30**, 1–3.

BRUMPT, E. (1912) 'Pénétration du *Schizotrypanum cruzi* à travers la muqueuse oculaire saine', *Bull. Soc. Path. exot.* **5**, 724–7.

BULLOCH, W. (1938) *The History of Bacteriology*, London: Oxford University Press.

BÜTTNER, D. W. (1967) 'Die Feinstruktur der Merozoiten von *Theileria parva*', *Z. Tropenmed. Parasit.* **18**, 224–44.

CANNING, E. O. and ANWAR, M. (1968) 'Studies on meiotic division in coccidial and malarial parasites', *J. Protozool.* **15**, 290–8.

CAROTHERS, J. C. (1934) 'An investigation of the etiology of subacute nephritis as seen among children in North Kavirondo', *E. Afr. med. J.* **10**, 335–6.

CHAGAS, C. (1909) 'Neue Trypanosomen: *T. minasense* n.sp., *T. cruzi*, n.sp.', *Archiv. Schiffs. tropen. Hyg.* **13**, 120–33.

CHAGAS, E. (1936) 'Commentarios sobre la vida e a obra di Carlos Chagas', *Reunion Soc. Argent. Patol. Reg.* **1**, 120–35.

CLYDE, D. (1967) *Malaria in Tanzania*, London: Oxford University Press.

COLE, F. J. (1926) *The History of Protozoology*, London: University of London Press.

COLUZZI, M. and TRABUCCHI, R. (1968) 'Importanza dell' armatura bucco-faringea in *Anopheles* e *Culex* in relazione alle infezioni con *Dirofilaria*', *Parassitologia* **10**, 47–59.

CORRADETTI, A. (1954) 'L'opera protozoologica di Battista Grassi', *Riv. parassit.* **15**, 1–12.

CORRADETTI, A., VEROLINI, F. and BUCCI, A. (1966) 'Resistenza a *Plasmodium berghei* da parte di ratti albini precedentemente immunizzati con *P. berghei*', *Parassitologia* **8**, 133–45.

CORTI, A. (1955) 'Nel centenario della nascita di Battista Grassi', *Studi Med. Biol.* **2**, 1–41.

CORTI, A. (1961) 'Battista Grassi, e la trasmissione della malaria', *Studia Ghisleriana*, Collana di Monografie dell Associazione Alunni del Collegio Ghislieri in Pavia.

COTRONEI, G. (1954) 'Battista Grassi', *Accad. Naz. Lincei* **351**, Quad. 33.

DANILEWSKY, B. (1889) *La Parasitologie Comparée du Sang*, Kharkov.

DOLMAN, C. E. (1969) 'Theobald Smith, 1859–1934, Life and Work', *New York State J. Med.* **69**, 2801–16.

EDINGTON, G. M. and GILLES, H. M. (1969) *Pathology in the Tropics*, London: Edward Arnold.

FINKELSTEIN, E. A. (1955) *Basil Jakovlevich Danilewsky*, Moscow: Acad. Nat. U.S.S.R.

FLAUBERT, G. (1821–1880) *Salammbo* in *Oeuvres*, vol. I, Paris: Bibliothèque de la Pléiade.

FOSTER, W D (1965) *A History of Parasitology*, Edinburgh. Livingstone.

FRIEDHOFF, K. and SCHOLTYSECK, E. (1968) 'Feinstrukturen von *Babesia ovis* (Piroplasmidea) in *Rhipicephalus bursa* (Ixodoidea): Transformation sphäroider Formen zu Vermiculaformen', *Z. f. Parasitenkunde*, **30**, 347–59.

GARNHAM, P. C. C. (1967) 'Reflections on Laveran, Marchiafava, Golgi, Koch and Danilewsky after sixty years', *Trans. R. Soc. trop. Med. Hyg.* **61**, 735–64.

GARNHAM, P. C. C., HEISCH, R. B. and MINTER, D. M. (1961) 'The vector of *Hepatocystis* (=*Plasmodium*) *kochi*; the successful conclusion of observations in many parts of tropical Africa', *Trans. R. Soc. trop. Med. Hyg.* **55**, 497–502.

GHALIOUNGHI, P. (1963) *Magic and Medical Science in Ancient Egypt*, London: Hodder & Stoughton.

GIGLIOLI, G. (1930) *Malarial Nephritis*, London: J. & A. Churchill.

GIROUD, P. (1961) 'Charles Nicolle (1866–1936)', *Bull. Acad. Nat. Med.* **145**, 714–22.

GOETHE, J. W. (1749–1832) *Faust*, Part 1, Trans. Swanwick, London: G. Bell (1919).

GOGOL, N. V. (1809–1852) 'The Fair at Sorotchintsi' in *Evenings near the village of Dikanka*, Moscow: Foreign Language Publishing House.

GOLGI, C. (1929) *Studi Sulla Malaria*, Rome: L. Pozzi.

GRASSI, B. (1900) 'Studi di un zoologo sul malaria', *Att. Accad. Lincei. Memoria* (2nd ed. 1901) **3**, 299–511.

GRASSI, B. (1901) 'A proposito del paludismo senza malaria', *Rend. Accad. Lincei* **10**, 2, 123–31.

GRASSI, B., BASTIANELLI, G. and BIGNAMI, A. (1898, 1899) 'Coltivazione della similune malariche del l'uomo del' *Anopheles claviger*', *Att. Accad. Lincei.* **7**, 313–14 and **8**, 21–28.

GUIMARÃES ROSA J. (1962) *The Devil to Pay in the Backlands*, New York: Alfred A. Knopf.

HAWTHORNE, N. (1804–1863) *The Marble Faun*, New York: Houghton and Mifflin.

HALL, M. C. (1935) 'Theobald Smith as a parasitologist', *J. Parasit.* **21**, 231–43.

HEINE, H. (1797–1856) *A Tour in the Harz*. Transl. Francis Storr. In *Prose and Poetry*. London: Dent.

HEWITT, R. (1940) *Bird Malaria*, Baltimore: Johns Hopkins.

HOARE, C. A. (1939) Obituary, Basil Danilewsky, *Trans. R. Soc. trop. Med. Hyg.* **33**, 271–3.

HOARE, C. A. (1965) Academician Eugene N. Pavlovsky, Obituary, *Nature* **208**, 1151–2.

HOEPPLI, R. (1959) *Parasites and Parasitic Infections in Early Medicine and Science*, Singapore: Univ. Malaya Press.

HOEPPLI, R. (1969) *Parasitic Diseases in Africa and the Western Hemisphere*, Basel: Acta trop. Suppl. 10.

KAHN, C. (1949) *Aus dem Leben Fritz Richard Schaudinns*, Stuttgart: Georg Thiem.

KÖBERLE, F. (1968) 'Chagas disease and Chagas syndromes: The pathology of American trypanosomiasis' in *Advances in Parasitology*, vol. 6, London: Academic Press.

KOSSEL, H. (1899) 'Über einen malariaähnlichen Blutparasiten bei Affen', *Z. Hyg. Infekt. Krankh.* **32**, 25–32.

LAFACE, L. (1954) 'Gli studi entomologici di Battista Grassi', *Riv. parassit.* **15**, 1–27.

LAGRANGE, E. (1938) *Robert Koch, Sa Vie et Son Oeuvre*, Paris: Legrand.

LANGERON, M. (1907) 'Fritz Schaudinn: Notice biographique' *Arch. Parasit.* 388–408.

LAVERAN, A. (1899) 'Les hématozoaires endoglobulaires (Haemocytozoa)', *Cinquantenaire Soc. Biol.* 124–33.

LAVERAN, A. (1917) *Leishmanioses*, Paris: Masson.

LAVERAN, A. and FRANCHINI, G. (1923) 'Expériences sur les flagelles de la punaise du chou', *Bull. Soc. Path. exot.* **16**, 319–22.

LAVERAN, A. and MESNIL, F. (1904) *Trypanosomes et Trypanosomiases*, Paris: Masson.

LEE, D. L., LONG, P. L., MILLARD, B. J. and BRADLEY, J. (1969) 'The fine structure and method of feeding of the tissue stages of *Histomonas meleagridis*', *Parasitology* **59**, 171–84.

MAGALHÃES, O. (1944) 'Un poco de la vida de Carlos Chagas', *Rev, Circ. med. Mendozo* **11**, 3–16.

MANSON, P. (1894) 'On the nature and significance of the crescentic and flagellate bodies in malarial blood', *Brit. med. J.* ii, 1306–8.

MANSON, P. (1896) 'Hypothesis as to the life history of the malarial parasites outside the human body', *Lancet* **ii**, 1715–18.

MANSON-BAHR, P. (1963) 'The story of Malaria' in *International Review of Tropical Medicine*, vol. 2, New York: Academic Press.

MANSON-BAHR, P. and ALCOCK, A. (1927) *The Life and Work of Sir Patrick Manson*, London: Cassell & Co.

MANZONI, A. (1785–1873) *I Promessi Sposi*, London: Folio Soc.

MCGHEE, B. (1969) *Host distribution of the trypanosomid genus, Phytomonas in North America*. In *Progress in Protozoology*, Leningrad: Nauka.

NEUTRA, M. and LEBLOND, C. P. (1969) 'The Golgi apparatus', *Scientific American*, 100–7.

NICOLLE, C. (1932) *Biologie de l'invention*, Paris: F. Algan.

NICOLLE, C. (1961) *Destin des maladies infectieuses*, with preface by G. Duhamel and introduction by P. Nicolle, Geneva: Alliance Culturelle du Libre.

NICOLLE, P. (1966) 'Centenaire de Charles Nicolle', *Rev. Hyg. et Méd. soc.* **14**, 371–82.

NICOLLE, C. and MANCEAUX, L. (1908) 'Sur une infection à corps de Leishman (ou organismes voisins) du Gondi', *Compt. rend. Acad. Sc.* **147**, 743 44

NUTTALL, H. F. (1924) 'Biographical notes bearing on Koch', *Parasitology* **16**, 214–38.

OLPP, G. (1932) 'Hervorrangende Tropen Ärzte im Wort', *Med. Bild.* 360–4.

PAUTRIZEL, R., RIVIÈRE, M., PRIORÉ, A. and BERLUREAU, F. (1966) *C. r. Acad. Sci.* Paris **263**, 579.

PAVLOVSKY, E. N. (1939) *Vestnik Akad. Nauk. U.S.S.R.* no. 10.

PAVLOVSKY, E. N. (1958) *Poetry, Science and Scientists*, Akad. Sci. U.S.S.R. Moscow, Leningrad.

PAZZINI, A. (1935) Obituary: Grassi, *Riv. Biol.* **19**, 126–49.

REDI, F. (1684) *Osservazioni intorno agli animali viventi che si trovano negli animali viventi*, Florence: P. Matini.

ROSS, R. (undated) *Memories of Sir Patrick Manson.*

ROSS, R. (1920) *The Revels of Orsera*, London: John Murray.

SACERDOTTI, C. (1926) 'Camille Golgi', *Pathologica* **18**, 54–65.

SCHAUDINN, F. (1899) Über den Generationswechsel der Coccidian und die neuere Malariaforschung' in *Sitz. ges. Naturf. Freunde*, Berlin.

SCHAUDINN, F. (1902) 'Studien über krankheitserregende Protozoen', *Arb. K. gesund.* **19**, 169–250.

SCHAUDINN, F. (1904) 'Generations- und Wirtswechsel bei Trypanosoma und Spirochaete', *Arb. K. gesund.* **20**, 387–439.

SCHAUDINN, F. and SIEDLECKI, M. (1897) 'Beiträge zur Kenntniss der Coccidien' in *Verh. Deutsch. Zoolog. Ges.* 192–203.

SERGENT, ED. and ET., and PARROT, L. (1929) *La Découverte de Laveran*, Paris: Masson.

SHORTT, H. E. and GARNHAM, P. C. C. (1948) 'The pre-erythrocytic development of *Plasmodium cynomolgi* and *Plasmodium vivax*', *Trans. R. Soc. trop. Med. Hyg.* **41**, 785–95.

SILVESTRI, F. (1925) 'In Onoranze a Battista Grassi', *Tip. de Senato* Rome 31–73.

SMITH, T. (1934) Letter to E. B. Krumbhaar, M.D., 11 October 1933, *J. Bact.* **27**, 19.

STENDHAL, (1783–1842) *The Charterhouse of Parma*, London: Penguin.

THOM, W. (1798–1848) 'The Blind Boy's Pranks' in *Oxford Book of English Verse*, Oxford University Press.

THOMSON, J. D. (1917) 'Notes on malaria', *J. R. Army med. Corps* **29**, 379–83.

DU TOIT, P. J. and JACKSON, C. (1936) 'The life and work of Sir Arnold Theiler', *J. Vet. Sci.* **7**, 134–86.

WENYON, C. M. (1935) 'Theobald Smith, M.D. Obituary', *Trans. R. Soc. trop. Med. Hyg.* **28**, 663–4.

WICHTERMAN, R. (1969) 'Some problems in studying the biological effects of ionizing radiations on Protozoa', *Progress in Protozoology. Third Internat. Cong. Protozoology* Leningrad 17–19.

WOODRUFF, A. W. (1965) 'Darwin's health in relation to his voyage in South America', *Br. med. J.* i, 745–50.

YOELI, M. (1967) 'Charles Nicolle and the Frontiers of Medicine', *New England J. Med.* **276**, 760–775.

ADDENDA

Page 50

After the introduction of the parasite into domestic birds, the epizootic spreads within the flock, and the feral connection may vanish, as has happened in the malaria of chickens in the environs of Colombo.

Page 68

Soper (1970) still shares his compatriot's opinion and asserts that eradication should not be deferred until the completion of the public health infrastructure.

SOPER, F. L. (1970) *Building the Health Bridge: Selections from the works of Fred L. Soper M.D.*, Bloomington: Indiana University Press.

Page 148

ASIA

As mentioned above (p. 131), there are various parasitological establishments in Asia, but I only want to describe briefly a few in Ceylon which I have had the opportunity of visiting and which are staffed largely by my own students. The Medical Research Institute of Colombo is run much on the lines of a Pasteur Institute, and has a good parasitological division under C. de S. Kulasiri; early in the century it housed Muriel Robertson, F.R.S. who worked on blood protozoa of reptiles, and Aldo Castellani who made his discoveries on human toxoplasmosis and yaws in a laboratory still in use today. The doyen of parasitologists now is A. S. Dissanaike, Dean of the Medical Faculty of Colombo and well known for his work on monkey malaria and filariasis; one or more members of his staff are always undergoing postgraduate training in England. A subsidiary school has recently been founded at Peradeniya where Mrs Nelson and Dr Gooneratne are actively engaged in research on protozoology and helminthology respectively. This highland station also possess fine new Institutes of Veterinary Science with facilities (as in the other laboratories) for field work throughout the island; the parasitology departments are under the charge of Professor Seniveratne and S. B. Dhanapala.

INDEX